The Classical Moment Problem

The
Classical
Moment
Problem

and

Some Related Questions in Analysis

N. I. Akhiezer

Translated by
N. Kemmer

DOVER PUBLICATIONS, INC.
Garden City, New York

Bibliographical Note

This Dover edition, first published in 2020, is an unabridged republication of the work originally published in 1965 by Oliver and Boyd Ltd., Edinburgh and London, as part of the University Mathematical Monographs series. The book was originally published in Russian in 1961.

Library of Congress Cataloging-in-Publication Data

Names: Akhiezer, N. I. (Naum Il'ich), 1901–1980, author.
Title: The classical moment problem and some related questions in analysis / Naum Il'yich Akhiezer ; translated by N. Kemmer.
Other titles: Klassicheskaia problema momentov i nekotorye voprosy analiza, sviazannye s neiu. English
Description: Garden City, New York : Dover Publications, Inc., 2020. | This Dover edition, first published in 2020, is an unabridged republication of the work originally published in 1965 by Oliver and Boyd Ltd., Edinburgh and London, as part of the University Mathematical Monographs series. | Includes bibliographical references and index. | Summary: "The moment problem in mathematics focuses on a measure within a sequence over a temporal period. Issues associated with the moment problem involve probability theory as a measure of mean, variance, etc., throughout a defined temporal period. This text provides a classic treatment of such issues that also involve linear algebra, probability theory, stochastic processes, quantum fields, signal processing, and more"—Provided by publisher.
Identifiers: LCCN 2019057391 | ISBN 9780486845555 (trade paperback)
Subjects: LCSH: Functional analysis.
Classification: LCC QA320 .A413 2020 | DDC 515/.7—dc23
LC record available at https://lccn.loc.gov/2019057391

Manufactured in the United States by LSC Communications
84555901
www.doverpublications.com

2 4 6 8 10 9 7 5 3 1
2020

CONTENTS

CHAPTER 4

INCLUSION OF THE POWER MOMENT PROBLEM IN
THE SPECTRAL THEORY OF OPERATORS

CHAPTER 5

TRIGONOMETRIC AND CONTINUOUS ANALOGUES

FOREWORD

THE term moment problem occurs for the first time in the work of T. Stieltjes of 1894–1895. However, one important problem related to the moment problem was posed and solved in a particular case as early as 1873 by P. L. Chebyshev. Throughout the later part of his life Chebyshev repeatedly returned to this problem. An exhaustive study of Chebyshev's problem and of various generalizations was performed by A. Markov.

At the end of Chapter 4 of his memoir on continued fractions Stieltjes[3] writes: " We shall give the name moment problem to the following problem: it is required to find the distribution of positive mass on the interval $[0, \infty)$, given the moments of order $k (k = 0, 1, 2...)$ of the distribution ". To give a distribution of positive mass on $[0, \infty)$ means to give a non-decreasing function $\sigma(u)$ $(u \geq 0)$ such that for any $\alpha \geq 0$ and $\beta > \alpha$ the increase $\sigma(\beta) - \sigma(\alpha)$ represents the mass in the interval $[\alpha, \beta)$. The total mass on the line $[0, \infty)$ may therefore be written in the form

$$\lim_{\beta \to \infty} [\sigma(\beta) - \sigma(0)] = \int_0^\infty d\sigma(u),$$

while the integrals

$$\int_0^\infty u \, d\sigma(u), \quad \int_0^\infty u^2 \, d\sigma(u)$$

represent the static moment of the mass distribution in question and its moment of inertia with respect to the point $u = 0$. Stieltjes gives the name of generalized moment of order k to the integral

$$\int_0^\infty u^k \, d\sigma(u).$$

Thus in Stieltjes' problem a certain sequence of numbers s_k $(k = 0, 1, 2...)$ is given and a non-decreasing function $\sigma(u)$ $(u \geq 0)$ is sought such that

$$\int_0^\infty u^k \, d\sigma(u) = s_k \quad (k = 0, 1, 2, ...). \qquad [I]$$

A precursor of this is the following problem on the " limiting values of integrals "† (Chebyshev[1], Markov[1-4]): let $f(u)$ be an unknown function subject to the single condition that it remains positive between the limits of integration; given the values of the integrals

$$\int_a^b f(u)\, du, \quad \int_a^b u f(u)\, du, \quad ..., \quad \int_a^b u^{n-1} f(u)\, du$$

it is required to determine the exact upper and lower limits of the integral

$$\int_a^v f(u)\, du$$

for any value of v in the interval (a, b)".

Naturally the mechanical interpretation of this problem did not escape Chebyshev who writes at the end of his note: " As an example of a problem that may be solved by this method I shall mention the following: Given the length, weight, position of mass centre and moment of inertia of a material straight line with an unknown density varying from one point to the next, it is required to find the narrowest possible limits for the weight of any segment of the line ".

However, Chebyshev saw his problem primarily as a way of obtaining a certain limit theorem in probability theory. He did not fully succeed in realizing his intentions. This was done by Markov.

In the statement of Stieltjes' problem the carrier of the mass is the semi-axis $[0, \infty)$. Instead one can take the carrier of the mass to be any given finite interval or any other point set which forms part of the semi-axis $[0, \infty)$; if one does this a supplementary requirement has to be introduced into Stieltjes' problem, namely that a certain part of the semi-axis must be free of mass. Conversely, in the " extended " moment problem the carrier of mass is the whole axis $(-\infty, \infty)$; and therefore equation [1] is replaced by the following:

$$s_k = \int_{-\infty}^{\infty} u^k\, d\sigma(u) \quad (k = 0, 1, 2, ...). \qquad ...[2]$$

Any moment problem represents, essentially, a series of questions of which the most important are the questions of solubility, of the number of solutions and of the construction of all solutions, if they exist. Two solutions are not taken to be distinct if their difference is a constant at all points at which it is continuous, and the moment problem is called *determinate* if it has a unique solution in this sense.

† This name is due to Chebyshev.

All these questions applied to equation [1] were investigated by Stieltjes with amazing artistry and beauty. In Stieltjes' investigation continued fractions of a certain special form play a basic part. It should be noted that continued fractions of a somewhat more general type also form the basic apparatus used by Chebyshev and by Markov in their work on limiting values of integrals.

The first solution and discussion of the extended moment problem [2] is due to H. Hamburger[3], after whom that problem is usually named. Like his famous predecessors Hamburger makes wide use of continued fractions. Almost simultaneously with Hamburger problem [2] was solved, and advanced in certain directions by use of other methods, by P. Nevanlinna[1], M. Riesz[1-4], E. Hellinger[1] and T. Carleman[1-2]. This work revealed important connections between the moment problem and many branches of analysis. Of particular importance are the connections with functional analysis, the theory of functions and the spectral theory of operators.

This book is devoted to the detailed investigation of the moment problem and to the discussion of its relation to the mathematical disciplines just listed. It is also devoted in part to certain generalizations and continuous analogues of the classical problem.

Many questions relating to this sphere of ideas could not find a place in the main text but are in one way or another presented in the sections labelled " Addenda and Problems ". These sections are not uniform, either in content or in form. They contain both easy exercises and complicated theorems, the proof of which occupies many pages. In some cases it proved necessary to limit the discussion to a formulation of the result and to references to the literature.

The main text of the book had its origin in lectures delivered by the author in Kharkov university in which certain remarks made by those attending proved extremely useful, as did the opinions of colleagues. Particular gratitude is due to I. M. Glazman, N. S. Landkof, B. Ya. Levin and I. V. Ostrovskii for reading the manuscript and for critical remarks. Thanks is also due to the publishing editor L. Ya. Tslaf for his extensive help.

CHAPTER 1

INFINITE JACOBI MATRICES AND THEIR
ASSOCIATED POLYNOMIALS

The infinite matrices that will occur in this work play an important part in many questions of analysis and algebra and in the theory of mechanical oscillations. They are also related to a particular type of algebraic continued fraction which has many applications in nineteenth-century classical work especially of P. L. Chebyshev. These fractions were in particular the means of introducing so-called orthogonal systems of polynomials, which form the essential algebraic apparatus of the subsequent development. The present chapter is essentially devoted to this algebraic apparatus.

1. Basic concepts

1.1. There exist two kinds of objects, each of which may be taken as the starting points of our immediate discussion. They are positive sequences on the one hand and $\mathcal{J}$-matrices on the other.

By the term *positive* sequence (sometimes, *positive relative to an axis*) we shall describe an infinite sequence of real numbers $\{s_k\}_0^\infty$ such that all quadratic (Hankel) forms

$$\sum_{i,\,k=0}^{m} s_{i+k} x_i x_k \quad (m = 0, 1, 2, \ldots) \qquad \ldots [1.1]$$

are positive definite. Thus the positive property of the sequence $\{s_k\}_0^\infty$ is equivalent to the fact that all the determinants

$$D_k = \begin{vmatrix} s_0 & s_1 & \ldots\ldots & s_k \\ s_1 & s_2 & \ldots\ldots & s_{k+1} \\ \cdot & \cdot & \cdot\cdot\cdot\cdot\cdot\cdot & \cdot \\ s_k & s_{k+1} & \cdots & s_{2k} \end{vmatrix} \quad (k = 0, 1, 2, \ldots)$$

1

are positive. By the term $\mathscr{J}$-matrix we shall denote an infinite Jacobi matrix

$$\left\|\begin{array}{ccccc} a_0 & b_0 & 0 & 0 & 0 \ldots \\ b_0 & a_1 & b_1 & 0 & 0 \ldots \\ 0 & b_1 & a_2 & b_2 & 0 \ldots \\ \cdot & \cdot & \cdot & \cdot & \cdot & \cdot \cdot \cdot \cdot \cdot \cdot \\ \cdot & \cdot & \cdot & \cdot & \cdot & \cdot \cdot \cdot \cdot \cdot \cdot \end{array}\right\| , \qquad \ldots [1.2]$$

in which all the a_k are real and all the b_k positive.

If any sequence of numbers $\{s_k\}_0^\infty$ (which is not necessarily positive) is given, then it is possible to define a functional S in the space of all polynomials by putting†

$$S\{R(\lambda)\} = p_0 s_0 + p_1 s_1 + \ldots + p_n s_n$$

where

$$R(\lambda) = p_0 + p_1 \lambda + \ldots + p_n \lambda^n.$$

This functional is evidently additive and homogeneous, i.e. linear in the algebraic sense. It is called *positive* if $R(u) \geqq 0$ ($-\infty < u < \infty$) and $R(u) \not\equiv 0$ implies $S\{R\} > 0$.

The positive character of the forms [1.1] is a necessary and sufficient condition for the functional S to be positive.‡

Indeed any polynomial $R(\lambda)$ that satisfies the condition $R(u) \geqq 0$ ($-\infty < u < \infty$), can be represented in the form‖

$$R(\lambda =) [A(\lambda)]^2 + [B(\lambda)]^2,$$

where each of the polynomials

$$A(\lambda) = \sum_0^m x_k \lambda^k, \quad B(\lambda) = \sum_0^m y_k \lambda^k$$

† Side by side with the notation $S\{R(\lambda)\}$ we shall also use in the following the notations $S\{R\}$, $S_\lambda\{R(\lambda)\}$, $S_u\{R(u)\}$.
‡ This circumstance justifies the term " positive sequence ".
‖ In order to obtain this representation we note that here the decomposition of $R(\lambda)$ into linear factors has the form

$$R(\lambda) = c \prod_{k=1}^m (\lambda - \alpha_k - i\beta_k)(\lambda - \alpha_k + i\beta_k),$$

where $c > 0$, $\beta_k \geqq 0$, and the α_k are real (of course some of the $\alpha_k + i\beta_k$ may be equal). Now one simply has to select half of the factors and write

$$\sqrt{c} \prod_{k=1}^m (\lambda - \alpha_k - i\beta_k) = A(\lambda) + iB(\lambda),$$

where $A(\lambda)$ and $B(\lambda)$ are polynomials with real coefficients.

has real coefficients. It follows from this representation that

$$\mathsf{S}\{R\} = \sum_{i,\,k=0}^{m} s_{i+k} x_i x_k + \sum_{i,\,k=0}^{m} s_{i+k} y_i y_k,$$

whence our statement is evident.

If the sequence $\{s_k\}_0^\infty$ is positive, then the sequence $\{\vartheta s_k\}_0^\infty$ has the same property for any $\vartheta > 0$. Taking into account also that for a positive sequence we have $s_0 > 0$, we can confine our attention to normalized positive sequences, by which we mean such sequences for which $s_0 = 1$.

1.2. Given a positive sequence $\{s_k\}_0^\infty$, it is possible to construct a sequence of polynomials

$$P_0(\lambda), \quad P_1(\lambda), \quad P_2(\lambda), \ldots,$$

which have the following two properties that characterize them fully:

(1) $P_n(\lambda)$ is a polynomial of exact degree n and its leading coefficient is positive.

(2) The following orthogonality relations hold:

$$\mathsf{S}\{P_m P_n\} = \delta_{mn} = \begin{cases} 0 & (m \neq n), \\ 1 & (m = n). \end{cases} \quad \ldots [1.3]$$

The polynomials $P_n(\lambda)$ $(n = 0, 1, 2, \ldots)$ are then called *orthogonal* and *normalized* (more briefly *orthonormal*) *with respect to the sequence*.

To obtain the polynomials $P_n(\lambda)$ one can use the well-known procedure of orthogonalization, applied to the sequence of functions λ^k $(k = 0, 1, 2, \ldots)$, taking as the scalar product the expression

$$(\lambda^i, \lambda^k) = s_{i+k}.$$

However one can also write down an explicit expression for these polynomials directly in terms of determinants, namely†

$$P_0(\lambda) = 1,$$

$$P_n(\lambda) = \frac{1}{\sqrt{D_{n-1} D_n}} \begin{vmatrix} s_0 & s_1 & \ldots & s_n \\ s_1 & s_2 & \ldots & s_{n+1} \\ \cdot & \cdot & \cdot & \cdot \\ s_{n-1} s_n & & \ldots & s_{2n-1} \\ 1 & \lambda & \ldots & \lambda^n \end{vmatrix} \quad (n = 1, 2, 3, \ldots). \quad \ldots [1.4]$$

It follows from [1.4] that

$$P_n(\lambda) = \sqrt{\frac{D_{n-1}}{D_n}} \lambda^n + R_{n-1}(\lambda),$$

† If one agrees to put $D_{-1} = 1$ the general determinantal expression remains valid for $n = 0$.

where $R_{n-1}(\lambda)$ is some polynomial of degree $n-1$. In order to prove therefore that the polynomials [1.4] satisfy the relations [1.3] one must verify the equations

$$S\{P_n(\lambda)\lambda^m\} = \begin{cases} 0 & (m = 0, 1, 2, \ldots, n-1), \\ \sqrt{\dfrac{D_n}{D_{n-1}}} & (m = n). \end{cases}$$

To do this it is sufficient to note the identity

$$P_n(\lambda)\lambda^m = \frac{1}{\sqrt{D_{n-1}D_n}} \begin{vmatrix} s_0 & s_1 & \cdots & s_n \\ s_1 & s_2 & \cdots & s_{n+1} \\ \cdots & \cdots & \cdots & \cdots \\ s_{n-1}s_n & \cdots & s_{2n-1} \\ \lambda^m & \lambda^{m+1} & \cdots & \lambda^{m+n} \end{vmatrix}$$

and to apply the functional S to both sides of it.

1.3. Any polynomial may be expanded in terms of the polynomials P_k. Therefore

$$\lambda P_k(\lambda) = a_{k,k+1}P_{k+1}(\lambda) + a_{k,k}P_k(\lambda) + a_{k,k-1}P_{k-1}(\lambda) + \ldots \quad \ldots [1.5]$$

Comparison of the leading coefficients gives:

$$a_{k,k+1} = \frac{\sqrt{D_{k-1}D_{k+1}}}{D_k}.$$

Multiplying both sides of [1.5] by $P_i(\lambda)$ $(i = 0, 1, \ldots, k)$ and applying the functional S we obtain:

$$\begin{aligned} a_{k,i} &= 0 \quad (i = 0, 1, 2, \ldots, k-2), \\ a_{k,k-1} &= S\{\lambda P_k(\lambda)P_{k-1}(\lambda)\}, \\ a_{k,k} &= S\{\lambda P_k(\lambda)P_k(\lambda)\}. \end{aligned}$$

Further, making the substitution

$$\lambda P_{k-1}(\lambda) = a_{k-1,k}P_k(\lambda) + R_{k-1}(\lambda)$$

in the second of these relations under the sign of the functional we find that

$$a_{k,k-1} = a_{k-1,k}.$$

Therefore the expansion [1.5] has the form

$$\lambda P_k(\lambda) = b_{k-1}P_{k-1}(\lambda) + a_kP_k(\lambda) + b_kP_{k+1}(\lambda) \quad (k = 0, 1, 2, \ldots),$$

where

$$b_{-1} = 0$$

and

$$a_k = a_{k,k} = S\{\lambda P_k(\lambda)P_k(\lambda)\}, \quad b_k = a_{k,k+1} = \frac{\sqrt{D_{k-1}D_{k+1}}}{D_k}.$$

In this way we have obtained a second order finite difference equation

$$b_{k-1}y_{k-1} + a_k y_k + b_k y_{k+1} = \lambda y_k \quad (k = 1, 2, 3, \ldots) \quad \ldots[1.6]$$

with the initial condition

$$(a_0 - \lambda)y_0 + b_0 y_1 = 0, \qquad \ldots[1.7]$$

from which we can successively find the y_k ($k = 1, 2, 3, \ldots$) if the value of y_0 is given.

Equation [1.6] and equation [1.7] determine a Jacobi matrix [1.2] of the form we are discussing.

1.4. Let us now make it our starting point that the $\mathscr{J}$-matrix [1.2] is given. In this case we can construct the finite difference equation [1.6] and take the initial condition [1.7] and then, putting $y_0 = 1 = P_0(\lambda)$, we can obtain successively the polynomials $y_k = P_k(\lambda)$, ($k = 1, 2, 3, \ldots$). An arbitrary polynomial can then be expressed in terms of these.

Now the functional S can be constructed. It must be linear and its domain of definition must be the whole of space. In the first place we assume

$$S\{P_i(\lambda)P_k(\lambda)\} = \delta_{ik},$$

i.e. we define the functional for certain polynomials of special form. By virtue of this definition and the linearity property we can attempt to define the functional for an arbitrary polynomial. Indeed any polynomial $R(\lambda)$ can be expressed in the form of a product of two polynomials, and if

$$R(\lambda) = A(\lambda)B(\lambda),$$

where

$$A(\lambda) = \sum_{i=0}^{m} A_i P_i(\lambda), \quad B(\lambda) = \sum_{k=0}^{n} B_k P_k(\lambda),$$

we have

$$S\{R(\lambda)\} = \sum_{i=0}^{m} \sum_{k=0}^{n} A_i B_k S\{P_i(\lambda)P_k(\lambda)\} = \sum_{i=0}^{r} A_i B_i \quad (r = \min\{m, n\}).$$

However, for such a definition to be admissible it is necessary to verify that the result to which it leads is independent of the way the polynomial $R(\lambda)$ is split into two factors. Evidently one only has to show that for arbitrary polynomials $M(\lambda)$ and $N(\lambda)$ the relation

$$S\{A_1(\lambda)B_1(\lambda)\} = S\{A_2(\lambda)B_2(\lambda)\},$$

holds where

$$A_1(\lambda) = \lambda M(\lambda), \quad B_1(\lambda) = N(\lambda),$$
$$A_2(\lambda) = M(\lambda), \quad B_2(\lambda) = \lambda N(\lambda).$$

By virtue of the linearity property it is sufficient to perform the proof for $M(\lambda) = P_m(\lambda)$, $N(\lambda) = P_n(\lambda)$. But in this case we have

$$A_1(\lambda) = \lambda P_m(\lambda) = b_m P_{m+1}(\lambda) + a_m P_m(\lambda) + b_{m-1} P_{m-1}(\lambda),$$
$$B_2(\lambda) = \lambda P_n(\lambda) = b_n P_{n+1}(\lambda) + a_n P_n(\lambda) + b_{n-1} P_{n-1}(\lambda),$$

and the required relation becomes trivial.

Having constructed the functional S one obtains the sequence $\{s_k\}_0^\infty$ automatically; indeed

$$s_k = S\{\lambda^k\} \quad (k = 0, 1, 2, \ldots).$$

We now prove that this sequence is positive. To do this we note that

$$\sum_{i,\,k=0}^{m} s_{i+k} x_i x_k = S\{[A(\lambda)]^2\},$$

where

$$A(\lambda) = \sum_0^m x_i \lambda^i.$$

The polynomial $A(\lambda)$ will now be expressed in terms of the polynomials $P_k(\lambda)$:

$$A(\lambda) = \sum_0^{\lceil m} \xi_k P_k(\lambda).$$

Hence

$$S\{[A(\lambda)]^2\} = \sum_0^m \xi_k^2.$$

Therefore

$$\sum_{i,\,k=0}^{m} s_{i+k} x_i x_k = \sum_0^m \xi_k^2, \qquad \ldots[1.8]$$

and this quantity is positive provided $A(\lambda)$ is not identically zero.

We see therefore that every $\mathscr{J}$-matrix generates a positive sequence.

Thus we have proved that the totality of all $\mathscr{J}$-matrices and the totality of all positive normalized sequences stand in one-to-one correspondence.

In the following we shall often use the relation $[1.8]$ in which the $(m+1)$-dimensional vectors

$$x = \{x_0, x_1, \ldots, x_m\} \quad \text{and} \quad \xi = \{\xi_0, \xi_1, \ldots, \xi_m\}$$

are related by the identity

$$\sum_0^m x_i \lambda^i = \sum_0^m \xi_k P_k(\lambda). \qquad \ldots [1.9]$$

In addition to [1.8] the following relation is also valid

$$\sum_{i,\,k=0}^m s_{i+k+1} x_i x_k = \sum_0^m a_k \xi_k^2 + 2 \sum_0^{m-1} b_k \xi_k \xi_{k+1}. \qquad \ldots [1.10]$$

Indeed we have

$$\sum_{i,\,k=0}^m s_{i+k+1} x_i x_k = \mathsf{S}\left\{\lambda\left[\sum_0^m x_i \lambda^i\right]^2\right\} = \mathsf{S}\left\{\lambda\left[\sum_0^m \xi_k P_k(\lambda)\right]^2\right\},$$

and the rest follows by use of equations [1.6], [1.7], and [1.3].

1.5. We shall call a $\mathscr{J}$-matrix bounded if

$$\sup |a_i| < \infty, \quad \sup b_i < \infty.$$

THEOREM 1.1.5. (1) *If a $\mathscr{J}$-matrix is bounded and if*

$$\sup |a_i| + 2 \sup b_i \leq R < \infty, \qquad \ldots [1.11]$$

then for any m the two forms

$$\sum_{i,\,k=0}^m (Rs_{i+k} \pm s_{i+k+1}) x_i x_k \qquad \ldots [1.12]$$

are non-negative.

(2) *If the forms [1.12] are non-negative for any m then the $\mathscr{J}$-matrix is bounded.*

For proof we write down the following identity, which follows from [1.8], and [1.10]:

$$\sum_{i,\,k=0}^m (Rs_{i+k} \pm s_{i+k+1}) x_i x_k = R \sum_0^m \xi_k^2 \pm \sum_0^m a_k \xi_k^2 \pm 2 \sum_0^{m-1} b_k \xi_k \xi_{k+1}.$$

Further we note that

$$2b_k |\xi_k \xi_{k+1}| \leq b_k \xi_k^2 + b_k \xi_{k+1}^2.$$

Therefore

$$\sum_{i,\,k=0}^m (Rs_{i+k} \pm s_{i+k+1}) x_i x_k \geq \sum_0^m \xi_k^2 (R \pm a_k - b_k - b_{k-1}),$$

and each term on the right hand side is ≥ 0, if the condition [1.11] is satisfied. Thus the statement (1) is proved.

If all forms $[1.12]$ are non-negative then by virtue of $[1.8]$ and $[1.10]$ any forms

$$\sum_0^m (R \pm a_k)\xi_k^2 \pm 2 \sum_0^{m-1} b_k \xi_k \xi_{k+1}$$

are non-negative for arbitrary m. In particular putting

$$\zeta_0 = \xi_1 = \dots = \xi_{i-1} = \xi_{i+2} = \xi_{i+3} = \dots = \xi_m = 0,$$

we find that for arbitrary real ξ_i, ξ_{i+1} we have

$$(R \pm a_i)\xi_i^2 \pm 2b_i \xi_i \xi_{i+1} + (R \pm a_{i+1})\xi_{i+1}^2 \geq 0.$$

Hence it follows that

$$R \pm a_i \geq 0, \quad R \pm a_{i+1} \geq 0$$

and

$$b_i^2 \leq R^2 \pm R(a_i + a_{i+1}) + a_i a_{i+1}.$$

Therefore

$$|a_i| \leq R, \quad |a_{i+1}| \leq R$$

and

$$b_i^2 \leq R^2 + a_i a_{i+1} \leq 2R^2,$$

i.e. statement (2) is also proved.

2. Properties of the polynomials associated with a Jacobi matrix

2.1. The finite difference equation

$$b_{k-1} y_{k-1} + a_k y_k + b_k y_{k+1} = \lambda y_k \quad (k = 1, 2, 3, \dots) \quad \dots[1.13]$$

has two linearly independent solutions. Above we introduced the solution $P_k(\lambda)$ which satisfied the initial conditions

$$P_0(\lambda) = 1, \quad P_1(\lambda) = \frac{\lambda - a_0}{b_0}. \qquad \dots[1.13a]$$

Now we shall introduce a second solution which we shall denote by $Q_k(\lambda)$ $(k = 0, 1, 2, \dots)$, which satisfies the initial conditions

$$Q_0(\lambda) = 0, \quad Q_1(\lambda) = \frac{1}{b_0}. \qquad \dots[1.13b]$$

Thus $Q_k(\lambda)$ is a polynomial of exact degree $k-1$. It is called a *polynomial of the second kind*, while the orthogonal polynomial $P_k(\lambda)$ is often called a *polynomial of the first kind*. It is not difficult to verify that

$$Q_k(\lambda) = S_u \left\{ \frac{P_k(\lambda) - P_k(u)}{\lambda - u} \right\}. \qquad \dots[1.14]$$

Indeed, this equality is obvious for $k = 0, 1$ and for higher values of k it follows from the fact that its right hand side satisfies equation [*1.13*]. Using this equation and the initial conditions [*1.13a*] and [*1.13b*] it is also easy to obtain the following results, which are analogous to results in the theory of linear differential equations.

(1) *The analogue of the Liouville-Ostrogradskii formula:*

$$P_{k-1}(\lambda)Q_k(\lambda) - P_k(\lambda)Q_{k-1}(\lambda) = \frac{1}{b_{k-1}} \quad (k = 1, 2, 3, \ldots) . \qquad \ldots[1.15]$$

(2) *The analogue of Green's formula:* Let y_k be a solution of [*1.13*] with the parameter λ and z_k a solution of the same equation with the parameter μ. Then we have

$$b_{n-1}(y_{n-1}z_n - y_n z_{n-1}) - b_{m-1}(y_{m-1}z_m - y_m z_{m-1})$$
$$= (\mu - \lambda) \sum_m^{n-1} y_k z_k. \qquad \ldots[1.16]$$

In particular, taking $y_k = P_k(\lambda)$ and $z_k = P_k(\mu)$ and $m = 1$ we obtain the so-called *Christoffel-Darboux formula*

$$(\mu - \lambda) \sum_0^{n-1} P_k(\lambda)P_k(\mu) = b_{n-1}\{P_{n-1}(\lambda)P_n(\mu) - P_n(\lambda)P_{n-1}(\mu)\}. \quad \ldots[1.17]$$

The sum

$$h_n(\lambda, \mu) = \sum_0^n P_k(\lambda)P_k(\mu)$$

is often called the *kernel polynomial of degree n*, belonging to the given system of orthogonal polynomials $\{P_k(\lambda)\}_0^\infty$ (or to the functional S). It follows from the obvious equation

$$S_\lambda\{h_n(\lambda, \mu)P_k(\lambda)\} = P_k(\mu) \quad (k = 0, 1, \ldots, n)$$

that for any polynomial $R(\lambda)$ of degree $\leq n$ we have

$$S_\lambda\{h_n(\lambda, \mu)R(\lambda)\} = R(\mu).$$

This relation completely determines the kernel polynomial. Hence, it follows incidentally that

$$h_n(\lambda, \mu) = -\frac{1}{D_n} \begin{vmatrix} 0 & 1 & \lambda & \ldots & \lambda^n \\ 1 & s_0 & s_1 & \ldots & s_n \\ \mu & s_1 & s_2 & \ldots & s_{n+1} \\ \ldots & \ldots & \ldots & \ldots & \ldots \\ \mu^n & s_n & s_{n+1} & \ldots & s_{2n} \end{vmatrix}.$$

Further, in the general formula $[1.16]$ we put

$$m = 1, \quad \mu = \bar{\lambda} \ (\operatorname{Im} \lambda \neq 0), \quad y_k = wP_k(\lambda) + Q_k(\lambda), \quad z_k = \bar{y}_k,$$

where w is a complex parameter. We then obtain the equation

$$\sum_0^{n-1} | \ wP_k(\lambda) + Q_k(\lambda) \ |^2 - \frac{w - \bar{w}}{\lambda - \bar{\lambda}}$$

$$= b_{n-1} | \ wP_{n-1}(\lambda) + Q_{n-1}(\lambda) \ |^2 \frac{\operatorname{Im} \dfrac{wP_n(\lambda) + Q_n(\lambda)}{wP_{n-1}(\lambda) + Q_{n-1}(\lambda)}}{\operatorname{Im} \lambda}. \quad ...[1.18]$$

2.2. Let us agree to use the term *quasiorthogonal* polynomial of degree n for the polynomial

$$\alpha P_n(\lambda) - \beta P_{n-1}(\lambda) = P_n(\lambda; \alpha, \beta),$$

where α, β are parameters which do not become 0 simultaneously. If $\alpha \neq 0$, we have

$$P_n(\lambda; \alpha, \beta) = \alpha P_n\left(\lambda, \frac{\beta}{\alpha}\right),$$

where

$$P_n(\lambda, \tau) = P_n(\lambda) - \tau P_{n-1}(\lambda).$$

It is easy to see that the following orthogonality relation holds:

$$\mathfrak{S}\{P_n(\lambda; \alpha, \beta)\lambda^k\} = 0 \quad (k = 0, 1, 2, ..., n-2).$$

THEOREM **1.2.2.** (1) *All the zeros of a real quasiorthogonal polynomial are real and simple.* (2) *Any two zeros of the polynomial $P_n(\lambda)$ are separated by a zero of the polynomial $P_{n-1}(\lambda)$ and vice versa.* (3) *The zeros of the polynomial $Q_n(\lambda)$ are real and simple and any two of them are separated by a zero of the polynomial $P_n(\lambda)$, and vice versa.*

Proof. (1) Assume that the real quasiorthogonal polynomial $P_n(\lambda, \tau)$, changes its sign only at the following points on the real axis:

$$\lambda_1 < \lambda_2 < ... < \lambda_m.$$

Then the polynomial

$$R(\lambda) = P_n(\lambda, \tau)(\lambda - \lambda_1)(\lambda - \lambda_2) ... (\lambda - \lambda_m) \not\equiv 0$$

is non-negative on the real axis and therefore $\mathfrak{S}\{R(\lambda)\} > 0$. But this contradicts the orthogonality relation if $m \leq n - 2$. Therefore $m \geq n - 1$ i.e. the polynomial $P_n(\lambda, \tau)$ has at least $n - 1$ simple real zeros. Therefore all its zeros are real and simple.

(2) It follows from the Christoffel-Darboux formula that for $\mu \to \lambda$ we have:

$$\sum_0^{n-1} [P_k(\lambda)]^2 = b_{n-1}\{P_{n-1}(\lambda)P_n'(\lambda) - P_n(\lambda)P_{n-1}'(\lambda)\}.$$

If λ_1 and λ_2 are two neighbouring zeros of the polynomial $P_n(\lambda)$, then the numbers $P_n'(\lambda_1)P_n'(\lambda_2)$ are of different sign and by virtue of the stated equality we have

$$b_{n-1}P_{n-1}(\lambda_1)P_n'(\lambda_1) = \sum_0^{n-1} [P_k(\lambda_1)]^2 > 0,$$

$$b_{n-1}P_{n-1}(\lambda_2)P_n'(\lambda_2) = \sum_0^{n-1} [P_k(\lambda_2)]^2 > 0.$$

Therefore the numbers $P_{n-1}(\lambda_1)$ and $P_{n-1}(\lambda_2)$ are of different sign, whence statement (2) follows.

(3) Statement 3 is proved analogously by using equation [1.15].

2.3. In subsequent discussions the following functions will play an important part:

$$w_n(\lambda, \tau) = -\frac{Q_n(\lambda) - \tau Q_{n-1}(\lambda)}{P_n(\lambda) - \tau P_{n-1}(\lambda)} = -\frac{Q_n(\lambda, \tau)}{P_n(\lambda, \tau)} \qquad \ldots[1.19]$$

They are functions of the complex variable λ, of the real parameter τ ($-\infty < \tau \leq \infty$) and of the index n. It follows from the definition of this function that

$$w_n(\lambda, \infty) = w_{n-1}(\lambda, 0).$$

The polynomial $Q_n(\lambda, \tau)$, is often called the *numerator corresponding to the quasiorthogonal polynomial* $P_n(\lambda, \tau)$.

THEOREM 1.2.3. (HELLINGER[1])† *let λ be fixed in the half-plane*

$$\text{Im } \lambda > 0 \ (\text{Im } \lambda < 0)$$

and let τ vary along the whole real axis. Then $w = w_n(\lambda, \tau)$ describes a circular contour $K_n(\lambda)$ in the half-plane $\text{Im } w > 0$ ($\text{Im } w < 0$); *the centre of this circle is at the point*

$$-\frac{Q_n(\lambda)\overline{P_{n-1}(\lambda)} - Q_{n-1}(\lambda)\overline{P_n(\lambda)}}{P_n(\lambda)\overline{P_{n-1}(\lambda)} - P_{n-1}(\lambda)\overline{P_n(\lambda)}}, \qquad \ldots[1.20]$$

† The circles $K_n(\lambda)$ are analogous to the circles first introduced by H. WEYL[1] in connection with the Sturm-Liouville equation on the semi-axis (i.e. in connection with the continuous analogue of equation [1.13]). Hellinger showed that other constructs given by Weyl can also be taken over from the Sturm-Liouville equation to equation [1.13].

Simultaneously with Hellinger the circles $K_n(\lambda)$ were found in a different way by R. NEVANLINNA[1] and somewhat later also by M. RIESZ[3]. These circles also occur in HAMBURGER'S[3] work but are introduced only to illustrate certain results.

and its radius is

$$\frac{1}{|\lambda-\bar{\lambda}|}\;\frac{1}{\sum\limits_0^{n-1}|P_k(\lambda)|^2}\,. \qquad \dots[1.21]$$

The equation of the circle $K_n(\lambda)$ may be written in the form

$$\frac{w-\bar{w}}{\lambda-\bar{\lambda}}-\sum_0^{n-1}|wP_k(\lambda)+Q_k(\lambda)|^2=0. \qquad \dots[1.22]$$

Proof. Since $w_n(\lambda,\tau)$ is a linear fraction in τ this function does indeed describe a certain circle when τ varies along the real axis. In order to find the centre and the radius of this circle we write down the identity

$$-\frac{Q_n(\lambda)-\tau Q_{n-1}(\lambda)}{P_n(\lambda)-\tau P_{n-1}(\lambda)}=-\frac{Q_n(\lambda)\overline{P_{n-1}(\lambda)}-Q_{n-1}(\lambda)\overline{P_n(\lambda)}}{P_n(\lambda)\overline{P_{n-1}(\lambda)}-P_{n-1}(\lambda)\overline{P_n(\lambda)}}$$

$$+\frac{Q_n(\lambda)P_{n-1}(\lambda)-Q_{n-1}(\lambda)P_n(\lambda)}{P_n(\lambda)\overline{P_{n-1}(\lambda)}-P_{n-1}(\lambda)\overline{P_n(\lambda)}}\frac{\overline{P_n(\lambda)}-\tau\overline{P_{n-1}(\lambda)}}{P_n(\lambda)-\tau P_{n-1}(\lambda)}\,.$$

Hence it follows that

$$w_n(\lambda,\tau)=-\frac{Q_n(\lambda)\overline{P_{n-1}(\lambda)}-Q_{n-1}(\lambda)\overline{P_n(\lambda)}}{P_n(\lambda)\overline{P_{n-1}(\lambda)}-P_{n-1}(\lambda)\overline{P_n(\lambda)}}$$

$$+\left|\frac{Q_n(\lambda)\overline{P_{n-1}(\lambda)}-Q_{n-1}(\lambda)\overline{P_n(\lambda)}}{P_n(\lambda)\overline{P_{n-1}(\lambda)}-P_{n-1}(\lambda)\overline{P_n(\lambda)}}\right|\cdot e^{i\theta},$$

where $\theta=\theta(\tau)$ is a real quantity. The above statement concerning the centre of the circle evidently follows, and for its radius one obtains the expression

$$\left|\frac{Q_n(\lambda)\overline{P_{n-1}(\lambda)}-Q_{n-1}(\lambda)\overline{P_n(\lambda)}}{P_n(\lambda)\overline{P_{n-1}(\lambda)}-P_{n-1}(\lambda)\overline{P_n(\lambda)}}\right|\,.$$

But by virtue of equations [1.15] and [1.17] we have

$$\frac{Q_n(\lambda)\overline{P_{n-1}(\lambda)}-Q_{n-1}(\lambda)\overline{P_n(\lambda)}}{P_n(\lambda)\overline{P_{n-1}(\lambda)}-P_{n-1}(\lambda)\overline{P_n(\lambda)}}=\frac{1}{(\lambda-\bar{\lambda})\sum\limits_0^{n-1}|P_k(\lambda)|^2},$$

and therefore the statement of the theorem concerning the radius is also proved.

We now take the identity [*1.18*] and insert into it the expression [*1.19*] in place of w. Since it follows from [*1.19*] that

$$\frac{w_n(\lambda, \tau)P_n(\lambda)+Q_n(\lambda)}{w_n(\lambda, \tau)P_{n-1}(\lambda)+Q_{n-1}(\lambda)} = \tau,$$

and since Im $\tau = 0$, the identity [*1.18*] takes on the form [*1.22*], which proves the last part of the theorem.

We shall use the notation $K_n(\lambda)$ both for the whole circular area and for the circular contour surrounding it.

We now prove that for points w lying outside the circle $K_n(\lambda)$ the inequality

$$\frac{w-\bar{w}}{\lambda-\bar{\lambda}} - \sum_0^{n-1} |\, wP_k(\lambda)+Q_k(\lambda)\,|^2 < 0$$

holds while for lines within the circle $K_n(\lambda)$ one has the inequality

$$\frac{w-\bar{w}}{\lambda-\bar{\lambda}} - \sum_0^{n-1} |\, wP_k(\lambda)+Q_k(\lambda)\,|^2 > 0.$$

Since the quantity

$$\frac{w-\bar{w}}{\lambda-\bar{\lambda}} - \sum_0^{n-1} |\, wP_k(\lambda)+Q_k(\lambda)\,|^2$$

is 0 only on the circumference of $K_n(\lambda)$ it must have one sign at points outside the circle and the other sign for points in the interior. But we have

$$\frac{w-\bar{w}}{\lambda-\bar{\lambda}} - \sum_0^{n-1} |\, wP_k(\lambda)+Q_k(\lambda)\,|^2 = A\,|\,w\,|^2+Bw+\bar{B}\bar{w}+C,$$

where

$$A = - \sum_0^{n-1} |\, P_k(\lambda)\,|^2 < 0.$$

For sufficiently large values of $|\,w\,|$ the expression is certainly negative. This proves our statement.

We note further that the circle $K_n(\lambda)$ lies entirely within the circle $K_{n-1}(\lambda)$ and the circumferences of the circles touch. Indeed if a point w lies on the circle $K_n(\lambda)$ then we have

$$\frac{w-\bar{w}}{\lambda-\bar{\lambda}} - \sum_0^{n-1} |\, wP_k(\lambda)+Q_k(\lambda)\,|^2 = 0.$$

Therefore

$$\frac{w-\bar{w}}{\lambda-\bar{\lambda}} - \sum_0^{n-2} |\, wP_k(\lambda)+Q_k(\lambda)\,|^2 \geqq 0$$

and thus the point w lies within the circle $K_{n-1}(\lambda)$ or on its circumference. The existence of a common point on the circumferences $K_n(\lambda)$ and $K_{n-1}(\lambda)$, therefore also the fact that these two circles touch, follows from the relation

$$w_n(\lambda, \infty) = w_{n-1}(\lambda, 0).$$

2.4. To conclude the present section we introduce four sequences of polynomials which we shall require in the following. They are defined by the equations:

$$A_n(z) = z \sum_0^{n-1} Q_k(0)Q_k(z),$$

$$B_n(z) = -1 + z \sum_0^{n-1} Q_k(0)P_k(z),$$

$$C_n(z) = 1 + z \sum_0^{n-1} P_k(0)Q_k(z),$$

$$D_n(z) = z \sum_0^{n-1} P_k(0)P_k(z).$$

With the aid of Green's formula [1.16] these polynomials can be expressed in a different form:

$$\left.\begin{aligned}
A_n(z) &= b_{n-1}\{Q_{n-1}(0)Q_n(z) - Q_n(0)Q_{n-1}(z)\}, \\
B_n(z) &= b_{n-1}\{Q_{n-1}(0)P_n(z) - Q_n(0)P_{n-1}(z)\}, \\
C_n(z) &= b_{n-1}\{P_{n-1}(0)Q_n(z) - P_n(0)Q_{n-1}(z)\}, \\
D_n(z) &= b_{n-1}\{P_{n-1}(0)P_n(z) - P_n(0)P_{n-1}(z)\}.
\end{aligned}\right\} \quad ...[1.23]$$

Thus $B_n(z)$ and $D_n(z)$ are real quasiorthogonal polynomials of degree n while $A_n(z)$ and $C_n(z)$ are the corresponding numerators.

Using the relations [1.23] and the Liouville-Ostrogradskii equations [1.15] it is easy to obtain the relations

$$\left.\begin{aligned}
P_n(z) &= Q_n(0)D_n(z) - P_n(0)B_n(z), \\
P_{n-1}(z) &= Q_{n-1}(0)D_n(z) - P_{n-1}(0)B_n(z), \\
Q_n(z) &= Q_n(0)C_n(z) - P_n(0)A_n(z), \\
Q_{n-1}(z) &= Q_{n-1}(0)C_n(z) - P_{n-1}(0)A_n(z).
\end{aligned}\right\} \quad ...[1.23a]$$

Inserting these expressions into [1.15] we find that

$$A_n(z)D_n(z) - B_n(z)C_n(z) = 1. \qquad ...[1.24]$$

We note that the function $w_n(\lambda, \tau)$ defined above by equation [*1.19*] can be expressed in the form

$$w_n(\lambda, \tau) \equiv w_n(\lambda) = -\frac{A_n(\lambda)t - C_n(\lambda)}{B_n(\lambda)t - D_n(\lambda)}, \qquad \dots [1.25]$$

where t is a real parameter which is a linear function of τ. Therefore the point [*1.25*] describes the circular contour $K_n(\lambda)$ when t traverses the real axis and Im $\lambda \neq 0$.

3. Theorems of invariance and analyticity

3.1. Let us take any non-real point λ and construct for it the sequence of circles $K_n(\lambda)$. Since $K_{n+1}(\lambda) \subset K_n(\lambda)$, there exists either a limiting circle or a limiting point $K_\infty(\lambda)$.

If w happens to be the limiting point, or any point of the limiting circle it follows from the remark made at the end of Subsection 2.3. that for any positive integer n the inequality

$$\sum_0^{n-1} | wP_k(\lambda) + Q_k(\lambda) |^2 < \frac{w - \overline{w}}{\lambda - \overline{\lambda}}$$

is valid. Therefore

$$\sum_0^\infty | wP_k(\lambda) + Q_k(\lambda) |^2 < \infty.$$

Since

$$y_k = wP_k(\lambda) + Q_k(\lambda) \quad (k = 0, 1, 2, \dots)$$

is a solution of equation [*1.26*] the first statement in the following proposition† is true:

THEOREM 1.3.1. *(a) For any non-real λ there exists at least one solution $\{y_k\}_0^\infty$ of the equations*

$$b_{k-1}y_{k-1} + a_k y_k + b_k y_{k+1} = \lambda y_k \quad (k = 1, 2, \dots), \qquad \dots [1.26]$$

for which

$$\sum_0^\infty | y_k |^2 < \infty,$$

or in other words a solution which belongs to l^2

(b) Every solution of this equation belongs to l^2 if and only if $K_\infty(\lambda)$ is a circle.

† This proposition is the finite-difference analogue of the well known theorem due to WEYL[1] concerning the Sturm-Liouville equation on the semi-axis (for this theorem see for instance LEVITAN[1]).

Proof of the second statement. If every solution of [*1.26*] belongs to l^2, then

$$\sum_0^\infty |\,P_k(\lambda)\,|^2 < \infty$$

and therefore $K_\infty(\lambda)$ is a circle. Conversely if $K_\infty(\lambda)$ is a circle, then we have

$$\sum_0^\infty |\,P_k(\lambda)\,|^2 < \infty\,;$$

and since in any case

$$\sum_0^\infty |\,wP_k(\lambda)+Q_k(\lambda)\,|^2 < \infty \quad [w \in K_\infty(\lambda)],$$

the series

$$\sum_0^\infty |\,Q_k(\lambda)\,|^2$$

also converges.

3.2. THEOREM **1.3.2.** (ON INVARIANCE)†. *If $K_\infty(\lambda)$ is a circle for some non-real λ then $K_\infty(\lambda)$ will be a circle for any non-real λ.*

Further, if the series

$$\sum_0^\infty |\,P_k(\lambda)\,|^2$$

converges at any non-real point λ, then it converges uniformly in every finite part of the complex λ-plane.

We begin by proving an auxiliary proposition.

LEMMA **1.3.2.** *If*

$$\sum_0^\infty |\,c_n\,|^2 < \infty, \quad \sum_{n=0}^\infty \sum_{k=0}^{n-1} |\,a_{nk}\,|^2 < \infty$$

and if η_n ($n = 0, 1, 2, ...$) is a solution of the equation‡

$$\eta_n = c_n+(\lambda-\lambda_0)\sum_{k=0}^{n-1} a_{nk}\eta_k \quad (n = 0, 1, 2, ...), \qquad ...[1.27]$$

† A theorem of this kind was first established by Weyl for the Sturm-Liouville equation on the semi-axis. For the present case of a Jacobi matrix the theorem was obtained by HELLINGER[1] and NEVANLINNA[1].

The theory of Jacobi matrices and also the theory of the Sturm-Liouville equation are contained in the general spectral theory of operators. In this context the theorem on invariance proves to be a particular case of a certain more general fact. However, related problems concerning equations in which the parameter λ is involved as a linear fraction, are not directly contained in the theory of linear operators. Such problems were discussed by NEVANLINNA[2] and by WEYL[2] in one of his last papers. The proof we give of the invariance theorem is based on this last paper of Weyl's. It is quite elementary and does not depend on general methods of operator theory.

‡ Note that in passing to the continuous problem this equation goes over into a Volterra integral equation.

then the series

$$\sum_{0}^{\infty} |\eta_n|^2 \qquad \qquad ...[1.28]$$

converges uniformly in every finite part of the λ-plane.

To prove this lemma we write down the " triangle " inequality

$$\left\{ \sum_{n=m}^{N} |\eta_n|^2 \right\}^{\frac{1}{2}} \leqq \left\{ \sum_{n=m}^{N} |c_n|^2 \right\}^{\frac{1}{2}} + |\lambda - \lambda_0| \left\{ \sum_{n=m}^{N} \left| \sum_{k=0}^{n-1} a_{nk}\eta_k \right|^2 \right\}^{\frac{1}{2}}.$$

Assume that in a given finite region of the λ-plane

$$\sup |\lambda - \lambda_0| = R.$$

We choose an arbitrary positive number $\varepsilon < 1$ and find an $m = m(\varepsilon, R)$ such that

$$\left\{ \sum_{n=m}^{\infty} \sum_{k=0}^{n-1} |a_{nk}|^2 \right\}^{\frac{1}{2}} \leqq \frac{\varepsilon}{R},$$

$$\left\{ \sum_{n=m}^{\infty} |c_n|^2 \right\}^{\frac{1}{2}} \leqq \varepsilon.$$

Such a choice of m is possible by the conditions of the lemma. We keep the chosen m fixed. Then

$$\left\{ \sum_{n=m}^{N} |\eta_n|^2 \right\}^{\frac{1}{2}} \leqq \varepsilon + |\lambda - \lambda_0| \cdot \left\{ \sum_{n=m}^{N} \sum_{k=0}^{n-1} |a_{nk}|^2 \sum_{i=0}^{n-1} |\eta_i|^2 \right\}^{\frac{1}{2}}$$

$$\leqq \varepsilon + R \cdot \frac{\varepsilon}{R} \left\{ \sum_{i=0}^{N} |\eta_i|^2 \right\}^{\frac{1}{2}},$$

whence

$$(1-\varepsilon)\left\{ \sum_{n=m}^{N} |\eta_n|^2 \right\}^{\frac{1}{2}} \leqq \varepsilon + \varepsilon \left\{ \sum_{i=0}^{m-1} |\eta_i|^2 \right\}^{\frac{1}{2}}. \qquad ...[1.29]$$

This inequality shows in the first place that the series $[1.28]$ converges everywhere in the chosen region of the λ-plane. Further, the right hand side of the inequality $[1.29]$ being a continuous function of λ, has a finite upper bound in this region. By virtue of the inequality $[1.29]$ the sum of the series $[1.28]$ also has a finite upper bound, say M^2, in the region. Thus it follows from $[1.29]$ that

$$\left\{ \sum_{n=m}^{N} |\eta_n|^2 \right\}^{\frac{1}{2}} \leqq \frac{\varepsilon(1+M)}{1-\varepsilon}.$$

Since $\varepsilon > 0$ can be taken to be arbitrarily small, and N to be arbitrarily large, the lemma is proved.

Now we can turn to the proof of Theorem **1.3.2.** We write down the following identities, which follow from Green's formula:

$$b_{n-1}\{P_{n-1}(\lambda_0)P_n(\lambda)-P_n(\lambda_0)P_{n-1}(\lambda)\} = (\lambda-\lambda_0)\sum_0^{n-1}P_k(\lambda_0)P_k(\lambda),$$

$$b_{n-1}\{Q_{n-1}(\lambda_0)P_n(\lambda)-Q_n(\lambda_0)P_{n-1}(\lambda)\} = -1+(\lambda-\lambda_0)\sum_0^{n-1}Q_k(\lambda_0)P_k(\lambda).$$

Multiplying the first equation by $Q_n(\lambda_0)$ and the second by $-P_n(\lambda_0)$ and adding, we obtain the following equation, after some simple transformations†:

$$P_n(\lambda) = P_n(\lambda_0)+(\lambda-\lambda_0)\sum_{k=0}^{n-1}\{P_k(\lambda_0)Q_n(\lambda_0)-P_n(\lambda_0)Q_k(\lambda_0)\}P_k(\lambda).$$

This relation allows one to determine the $P_n(\lambda)$ $(n = 0, 1, 2, ...)$ consecutively if the $P_n(\lambda_0)$ and $Q_n(\lambda_0)$ $(n = 0, 1, 2, ...)$ are known. Putting

$$\left.\begin{array}{l}\eta_n = P_n(\lambda), \quad c_n = P_n(\lambda_0),\\ a_{nk} = P_k(\lambda_0)Q_n(\lambda_0)-P_n(\lambda_0)Q_k(\lambda_0),\end{array}\right\}\qquad ...[1.30]$$

we obtain equation [*1.27*]. Since $K_\infty(\lambda_0)$ is a circle, we have

$$\sum_0^\infty|P_n(\lambda_0)|^2 < \infty,$$

$$\sum_0^\infty|Q_n(\lambda_0)|^2 < \infty$$

† This equation can be derived more simply. Indeed the ratio
$$\frac{P_n(\lambda)-P_n(\lambda_0)}{\lambda-\lambda_0}$$
is a polynomial of degree $n-1$ in λ and therefore can be represented in the form
$$\sum_{k=0}^{n-1}a_{nk}P_k(\lambda).$$
The coefficients a_{nk} can be calculated according to the formula
$$a_{nk}=S_\lambda\left\{\frac{P_n(\lambda)-P_n(\lambda_0)}{\lambda-\lambda_0}P_k(\lambda)\right\}=P_k(\lambda_0)S_\lambda\left\{\frac{P_n(\lambda)-P_n(\lambda_0)}{\lambda-\lambda_0}\right\}+$$
$$+S_\lambda\left\{[P_n(\lambda)-P_n(\lambda_0)]\frac{P_k(\lambda)-P_k(\lambda_0)}{\lambda-\lambda_0}\right\};$$
it remains to note that
$$S_\lambda\left\{\frac{P_k(\lambda)-P_k(\lambda_0)}{\lambda-\lambda_0}\right\}=Q_k(\lambda_0)$$
and
$$S_\lambda\left\{P_n(\lambda)\frac{P_k(\lambda)-P_k(\lambda_0)}{\lambda-\lambda_0}\right\}=0 \quad (k=0, 1, 2, ..., n).$$
There is a certain advantage in using the proof given in the text, namely that it can also be applied to the continuous case.

and therefore the first condition of Lemma **1.3.2.** is certainly satisfied. But the second condition is also satisfied because equation [*1.30*] leads to the inequality

$$\sum_{n=0}^{\infty} \sum_{k=0}^{n-1} |a_{nk}|^2$$

$$\leqq 2\left\{ \sum_{n=0}^{\infty} |Q_n(\lambda_0)|^2 \sum_{k=0}^{n-1} |P_k(\lambda_0)|^2 + \sum_{n=0}^{\infty} |P_n(\lambda_0)|^2 \sum_{k=0}^{n-1} |Q_k(\lambda_0)|^2 \right\}$$

$$< 4 \sum_{n=0}^{\infty} |Q_n(\lambda_0)|^2 \sum_{k=0}^{\infty} |P_k(\lambda_0)|^2.$$

Therefore Lemma **1.3.2.** is applicable and so the series

$$\sum_{0}^{\infty} |P_n(\lambda)|^2$$

converges uniformly in any finite part of the λ-plane.

The series

$$\sum_{0}^{\infty} |Q_n(\lambda)|^2$$

has the same property; this may be proved analogously.

Having proved the invariance theorem one can speak of the alternative we have been discussing (that there is either a limiting circle or a limiting point) as of a property of the $\mathscr{J}$-matrix as a whole.

DEFINITION **1.3.2.** A $\mathscr{J}$-matrix is called a matrix of type C if we have the case of a limiting circle, and of type D if we have the case of a limiting point.

COROLLARY **1.3.2.** For a $\mathscr{J}$-matrix of type C the radius of the circle $K_\infty(\lambda)$ is a continuous function of λ in each of the half-planes Im $\lambda > 0$, and Im $\lambda < 0$.

3.3. We now take a $\mathscr{J}$-matrix of type D, so that there exists for every non-real λ a unique value $w = w(\lambda)$ for which

$$\sum_{0}^{\infty} |wP_k(\lambda) + Q_k(\lambda)|^2 < \infty.$$

What can be said of the function $w(\lambda)$?

THEOREM **1.3.3.** (CONCERNING ANALYTICITY). *A function $w(\lambda)$ which belongs to a $\mathscr{J}$-matrix of type D is regular in the half-plane* Im $\lambda > 0$ *and also in the half-plane* Im $\lambda < 0$. *We also have*

$$\frac{\text{Im } w(\lambda)}{\text{Im } \lambda} > 0.$$

Proof. In each of the half-planes $\operatorname{Im} \lambda > 0$ and $\operatorname{Im} \lambda < 0$, the function $w(\lambda)$ is the limit of a sequence of regular functions

$$w_n(\lambda, 0) = -\frac{Q_n(\lambda)}{P_n(\lambda)},$$

for which

$$\frac{w_n(\lambda, 0) - \overline{w_n(\lambda, 0)}}{\lambda - \bar{\lambda}} - \sum_{k=0}^{n-1} \mid w_n(\lambda, 0)P_k(\lambda) + Q_k(\lambda) \mid^2 = 0. \quad \dots[1.31]$$

Thus, in order to prove the theorem one only has to establish the fact that the set of functions $w_n(\lambda, 0)$ is uniformly bounded in any closed region which has no points in common with the real axis. But by virtue of the relation $[1.31]$ we have the inequality

$$\mid w_n(\lambda, 0) \mid^2 = \mid w_n(\lambda, 0)P_0(\lambda) + Q_0(\lambda) \mid^2 \leq \frac{w_n(\lambda, 0) - \overline{w_n(\lambda, 0)}}{\lambda - \bar{\lambda}},$$

whence

$$\mid w_n(\lambda, 0) \mid \leq \frac{2}{\mid \lambda - \bar{\lambda} \mid},$$

which shows that the required condition of boundedness is satisfied.

Going to the limit in $[1.31]$ we find that

$$\frac{w(\lambda) - \overline{w(\lambda)}}{\lambda - \bar{\lambda}} = \sum_0^\infty \mid w(\lambda)P_k(\lambda) + Q_k(\lambda) \mid^2 > 0,$$

which proves the second part of the theorem.

4. Quadrature formula. Continued fractions

4.1. We take a quasiorthogonal polynomial of degree n

$$P_n(\lambda, \tau) = P_n(\lambda) - \tau P_{n-1}(\lambda),$$

where τ is a real number and we denote the zeros of the polynomial by

$$\lambda_1 < \lambda_2 < \dots < \lambda_n \quad [\lambda_k = \lambda_k^{(n)}(\tau)].$$

Assume further that $R_{2n-2}(\lambda)$ is an arbitrary polynomial of degree $2n - 2$. Then

$$R_{2n-2}(\lambda) = P_n(\lambda, \tau)R_{n-2}(\lambda) + R_{n-1}(\lambda),$$

where $R_{n-2}(\lambda)$ and $R_{n-1}(\lambda)$ are some polynomials of degrees $\leq n - 2$

and $\leqq n-1$ respectively. By Lagrange's interpolation formula we have

$$R_{n-1}(\lambda) = P_n(\lambda, \tau) \sum_{k=1}^n \frac{R_{n-1}(\lambda_k)}{P_n'(\lambda_k, \tau)(\lambda - \lambda_k)}$$

$$= P_n(\lambda, \tau) \sum_{k=1}^n \frac{R_{2n-2}(\lambda_k)}{P_n'(\lambda_k, \tau)(\lambda - \lambda_k)} \, .$$

On the other hand, from the orthogonality property of quasiorthogonal polynomials we have

$$\mathfrak{S}\{R_{2n-2}(\lambda)\} = \mathfrak{S}\{R_{n-1}(\lambda)\}.$$

Therefore

$$\mathfrak{S}\{R_{2n-2}(\lambda)\} = \sum_{k=1}^n \frac{R_{2n-2}(\lambda_k)}{P_n'(\lambda_k, \tau)} \mathfrak{S}_\lambda \left\{ \frac{P_n(\lambda, \tau)}{\lambda - \lambda_k} \right\} .$$

Since λ_k is a zero of the polynomial $P_n(\lambda, \tau)$ we have

$$\mathfrak{S}_\lambda \left\{ \frac{P_n(\lambda, \tau)}{\lambda - \lambda_k} \right\} = \mathfrak{S}_\lambda \left\{ \frac{P_n(\lambda, \tau) - P_n(\lambda_k, \tau)}{\lambda - \lambda_k} \right\}$$

$$= Q_n(\lambda_k) - \tau Q_{n-1}(\lambda_k) = Q_n(\lambda_k, \tau).$$

Therefore, for an arbitrary polynomial $R(\lambda)$ of degree $\leqq 2n-2$ the equality

$$\mathfrak{S}\{R(\lambda)\} = \sum_1^n \mu_k R(\lambda_k), \qquad \qquad \dots [1.32]$$

holds, where

$$\mu_k = \mu_k^{(n)}(\tau) = \frac{Q_n(\lambda_k, \tau)}{P_n'(\lambda_k, \tau)} . \qquad \qquad \dots [1.33a]$$

Formula $[1.32]$ is called the quadrature formula. For $\tau = 0$, i.e. for $P_n(\lambda, \tau) = P_n(\lambda)$ it is valid for any polynomial $R(\lambda)$, not only of degree $\leqq 2n-2$ but also of degree $2n-1$.

In addition to $[1.33a]$ the coefficients μ_k have two other important representations.

The first is obtained from $[1.33a]$ by using the fact that

$$P_n(\lambda_k) - \tau P_{n-1}(\lambda_k) = 0.$$

With the aid of this equation we obtain

$$\mu_k = \frac{P_{n-1}(\lambda_k)Q_n(\lambda_k) - P_n(\lambda_k)Q_{n-1}(\lambda_k)}{P_{n-1}(\lambda_k)P_n'(\lambda_k) - P_n(\lambda_k)P_{n-1}'(\lambda_k)}.$$

The numerator of the right hand side is equal to $1/b_{n-1}$. Further, using the Christoffel-Darboux formula we find that

$$\mu_k = \frac{1}{\sum_{i=0}^{n-1} |P_i(\lambda_k)|^2}, \qquad \qquad ...[1.33b]$$

whence we see incidentally that all the coefficients μ_k are positive. As one of the consequences of this fact we note the alternation† of the zeros of the polynomials $Q_n(\lambda, \tau)$ and $P_n(\lambda, \tau)$ for any real τ.

To obtain the second representation, we insert into the quadrature formula [1.32] the expression

$$R(\lambda) = \left[\frac{P_n(\lambda, \tau)}{P_n'(\lambda_k, \tau)(\lambda - \lambda_k)} \right]^2.$$

$R(\lambda)$ is a polynomial of degree $2n-2$ which is equal to 1 for $\lambda = \lambda_k$ and to 0 at the zeros of $P_n(\lambda, \tau)$. We then obtain the representation

$$\mu_k = S_\lambda \left\{ \left[\frac{P_n(\lambda, \tau)}{P_n'(\lambda_k, \tau)(\lambda - \lambda_k)} \right]^2 \right\}, \qquad ...[1.33c]$$

whence the positive nature of the μ_k follows again by virtue of the positive nature of the functional S.

4.2. We now prove that for any real finite τ the expansion

$$\frac{Q_n(z, \tau)}{P_n(z, \tau)} = \frac{s_0}{z} + \frac{s_1}{z^2} + \ ... \ + \frac{s_{2n-2}}{z^{2n-1}} + O\left(\frac{1}{z^{2n}}\right), \qquad ...[1.34a]$$

holds, and also that for $\tau = 0$ this expansion can be replaced by the following:

$$\frac{Q_n(z)}{P_n(z)} = \frac{s_0}{z} + \frac{s_1}{z^2} + \ ... \ + \frac{s_{2n-1}}{z^{2n}} + O\left(\frac{1}{z^{2n+1}}\right). \qquad ...[1.34b]$$

Indeed, by virtue of Lagrange's interpolation formula and of equation [1.33a] we have

$$\frac{Q_n(z, \tau)}{P_n(z, \tau)} = \sum_{k=1}^{n} \frac{Q_n(\lambda_k, \tau)}{P_n'(\lambda_k, \tau)(z - \lambda_k)} = \sum_{k=1}^{n} \frac{\mu_k}{z - \lambda_k}.$$

On the other hand we have

$$\sum_{k=1}^{n} \frac{\mu_k}{z - \lambda_k} = \frac{1}{z} \sum_{k=1}^{n} \mu_k + \frac{1}{z^2} \sum_{k=1}^{n} \mu_k \lambda_k + \ ... \ + \frac{1}{z^{m+1}} \sum_{k=1}^{n} \mu_k \lambda_k^m + \ ...$$

† This is a generalization of statement 3 of Theorem **1.2.2**.

and in order to prove [1.34a] it remains to note that

$$\sum_{k=1}^{n} \mu_k \lambda_k^m = S\{\lambda^m\} = s_m \quad (m = 0, 1, 2, \ldots, 2n-2).$$

If $\tau = 0$ the last relation is valid also for $m = 2n-1$, so that [1.34b] is also proved.

In the foregoing we have established a correspondence between normalized positive sequences and $\mathcal{J}$-matrices. Now we can describe this correspondence in different terms. To do this we introduce, in place of a positive sequence, the formal power series

$$\frac{s_0}{z} + \frac{s_1}{z^2} + \ldots + \frac{s_m}{z^{m+1}} + \ldots \quad (s_0 = 1), \qquad \ldots[1.35]$$

and in place of the $\mathcal{J}$-matrix the formal infinite continued fraction

$$\cfrac{1}{z - a_0 - \cfrac{b_0^2}{z - a_1 - \cfrac{b_1^2}{z - a_2 - \cdot}}}. \qquad \ldots[1.36]$$

We denote by

$$N_n(z) = z^{n-1} + \beta_1 z^{n-2} + \ldots + \beta_{n-1}$$

and by

$$M_n(z) = z^n + a_1 z^{n-1} + \ldots + a_n$$

the numerators and denominators of the consecutive finite approximations to the continued fraction [1.36] so that

$$\frac{N_1(z)}{M_1(z)} = \frac{1}{z - a_0}, \quad \frac{N_2(z)}{M_2(z)} = \cfrac{1}{z - a_0 - \cfrac{b_0^2}{z - a_1}}, \quad \ldots.$$

As is well known, and may also be verified very simply, the numerators and denominators of these fractions satisfy the finite-difference equation

$$Y_{k+1} = (z - a_k) Y_k - b_{k-1}^2 Y_{k-1} \quad (k = 1, 2, 3, \ldots),$$

provided one puts

$$\left.\begin{array}{ll} N_0(z) = 0, & N_1(z) = 1, \\ M_0(z) = 1, & M_1(z) = z - a_0. \end{array}\right\} \qquad \ldots[1.37]$$

We make the substitution

$$y_0 = Y_0, \quad y_k = \frac{1}{b_0 b_1 \ldots b_{k-1}} Y_k \quad (k = 1, 2, 3, \ldots).$$

We then obtain for y_k the equation

$$b_{k-1} y_{k-1} + a_k y_k + b_k y_{k+1} = z y_k \quad (k = 1, 2, 3, \ldots),$$

but this is just the finite-difference equation which we introduced in Section 1. Taking into account the initial conditions $[1.37]$ we therefore conclude that

$$M_0(z) = P_0(z), \quad N_0(z) = Q_0(z),$$

$$M_k(z) = b_0 b_1 \ldots b_{k-1} P_k(z), \quad N_k(z) = b_0 b_1 \ldots b_{k-1} Q_k(z)$$

$$(k = 1, 2, 3, \ldots).$$

We see that the approximating fractions of the continued fraction $[1.36]$ are given by

$$\frac{Q_k(z)}{P_k(z)}.$$

The formulae $[1.34b]$ show that the correspondence between the continued fraction $[1.36]$ and the power series $[1.35]$ consists of the fact that the expansion of the approximating fractions of degree n into a series of descending powers of z is the same as the series $[1.35]$ up to the term s_{2n-1}/z^{2n} inclusive. This correspondence determines uniquely the series $[1.35]$ in terms of the continued fraction $[1.36]$ and the continued fraction $[1.36]$ in terms of the series $[1.35]$. It played an important part in the work of Chebyshev and of Markov whose starting point was always the continued fractions. This is the formal correspondence which we had in mind in earlier remarks. In the following we shall encounter constructions which also lead to the continued fraction $[1.36]$, but which are based on different considerations.

Addenda and Problems

1. If

$$\sum_0^\infty \frac{1}{b_n} = \infty,$$

then the $\mathscr{J}$-matrix is of type D (CARLEMAN[3]).

Hint: use the equation

$$P_n(\lambda) Q_{n+1}(\lambda) - P_{n+1}(\lambda) Q_n(\lambda) = \frac{1}{b_n}.$$

2. If

$$\sum_0^\infty \frac{|a_{n+1}|}{b_n b_{n+1}} = \infty,$$

then the $\mathscr{J}$-matrix is of type D (DENNIS and WALL[1]).

Hint: use the equation

$$P_n(\lambda)Q_{n+2}(\lambda) - P_{n+2}(\lambda)Q_n(\lambda) = \frac{\lambda - a_{n+1}}{b_n b_{n+1}}.$$

3. If a finite constant M exists such that at least one of the in-equalities

$$b_{k-1} + a_k + b_k < M, \tag{α}$$

$$b_{k-1} - a_k + b_k < M \tag{β}$$

holds for all $k \geqq 1$, then the $\mathscr{J}$-matrix is of type D (WOUK[1]).

This theorem is the analogue of a theorem by Weyl on differential operators.

Proof. In case (α) we take the equation

$$b_{k-1}y_{k-1} + a_k y_k + b_k y_{k+1} = My_k \quad (k = 1, 2, 3, \ldots)$$

and rewrite it in the form

$$b_k(y_{k+1} - y_k) - b_{k-1}(y_k - y_{k-1}) = (M - b_{k-1} - a_k - b_k)y_k.$$

By virtue of (α) it follows that

$$\frac{1}{b_0} = Q_1(M) < Q_2(M) < \cdots,$$

and therefore the series

$$\sum_1^\infty |Q_n(M)|^2$$

diverges.

In the case (β) we take the equation

$$b_{k-1}z_{k-1} - a_k z_k + b_k z_{k+1} = Mz_k \quad (k = 1, 2, 3, \ldots),$$

which is satisfied by $(-1)^{k-1}Q_k(-M)$ and in analogy with the preceding argument we obtain the inequalities

$$\frac{1}{b_0} = Q_1(-M) < |Q_2(-M)| < \cdots.$$

4. If

$$\sum_0^\infty \frac{1}{\sqrt{b_n}} = \infty \tag{α}$$

and if there exists such a finite constant M that for any n and any real $\xi_0, \xi_1, \xi_2, \ldots \xi_n$ we have

$$\sum_0^n a_k \xi_k^2 + 2 \sum_0^{n-1} b_k \xi_k \xi_{k+1} < M \sum_0^n \xi_k^2, \qquad (\beta)$$

then the $\mathscr{J}$-matrix is of type D (WOUK[1]).

This theorem is the analogue of a theorem due to Hartman relating to differential operators. Condition (β) expresses the uniform semi-boundedness of the truncated matrices.

Proof. We begin by verifying that by virtue of (β) the inequalities

$$P_k(M) > 0, \quad Q_k(M) > 0 \quad (k = 1, 2, 3, \ldots).$$

hold. Then, with the aid of these inequalities and the relation

$$\frac{P_k(M)}{Q_k(M)} - \frac{P_{k+1}(M)}{Q_{k+1}(M)} = \frac{1}{b_k Q_k(M) Q_{k+1}(M)}$$

we establish the convergence of the series

$$\sum_1^\infty \frac{1}{b_k Q_k(M) Q_{k+1}(M)}.$$

After this it remains to apply the Cauchy inequality to the sum

$$\sum_1^n \frac{1}{\sqrt{b_k}} = \sum_1^n \frac{1}{\sqrt{b_k Q_k(M) Q_{k+1}(M)}} \sqrt{Q_k(M) Q_{k+1}(M)}$$

and to take into account condition (α).

5. If

$$|a_k| \leqq M < \infty \quad (k = 1, 2, 3, \ldots), \qquad (\alpha)$$

$$\sum_0^\infty \frac{1}{b_n} < \infty \qquad (\beta)$$

and if for all n, beginning with a certain value, we have

$$b_{n-1} b_{n+1} \leqq b_n^2, \qquad (\gamma)$$

then the $\mathscr{J}$-matrix is of type C.

(This is a particular case of a theorem due to BEREZANSKII[1].)

Proof. From the relation

$$b_{n-1} P_{n-1}(\lambda) + a_n P_n(\lambda) + b_n P_{n+1}(\lambda) = \lambda P_n(\lambda)$$

we find for $\lambda = i$ that

$$\sqrt{b_{n+1}}\,|\,P_{n+1}(i)\,| \leqq \frac{(1+|\,a_n\,|)\sqrt{b_{n+1}}}{b_n\sqrt{b_n}}\sqrt{b_n}\,|\,P_n(i)\,|$$

$$+\frac{\sqrt{b_{n-1}b_{n+1}}}{b_n}\sqrt{b_{n-1}}\,|\,P_{n-1}(i)\,|.$$

Hence by virtue of (α) and (γ) we find, starting from a certain n, that

$$\sqrt{b_{n+1}}\,|\,P_{n+1}(i)\,| \leqq \frac{1+M}{\sqrt{b_{n-1}b_n}}\sqrt{b_n}\,|\,P_n(i)\,|+\sqrt{b_{n-1}}\,|\,P_{n-1}(i).$$

Therefore the quantity

$$N_n = \max_{k=0,\,1,\,...,\,n}\sqrt{b_k}\,|\,P_k(i)\,|$$

satisfies the inequality

$$N_{n+1} \leqq N_n\left(1+\frac{1+M}{\sqrt{b_{n-1}b_n}}\right).$$

Taking into account (β) we conclude that

$$N_n \leqq N_1\prod_{k=1}^{\infty}\left(1+\frac{1+M}{\sqrt{b_{k-1}b_k}}\right) = N < \infty \quad (n = 1, 2, ...).$$

Therefore

$$\sqrt{b_n}\,|\,P_n(i)\,| \leqq N \quad (n = 0, 1, 2, ...)$$

and so by virtue of (β) we have

$$\sum_{n=0}^{\infty}|\,P_n(i)\,|^2 < \infty.$$

6. Prove that

$$\sum_{1}^{n}[Q_k(0)]^2 = -\frac{1}{D_n}\begin{vmatrix} 0 & 0 & s_0 & s_1 & \cdots & s_{n-1} \\ 0 & s_0 & s_1 & s_2 & \cdots & s_n \\ s_0 & s_1 & s_2 & s_3 & \cdots & s_{n+1} \\ \cdots\cdots\cdots\cdots\cdots\cdots \\ s_{n-1} & s_n & s_{n+1} & s_{n+2} & \cdots & s_{2n} \end{vmatrix}.$$

7. Assume that the $\mathscr{J}$-matrix

$$
I = \left\| \begin{array}{cccccc}
a_0 & b_0 & 0 & 0 & 0 \ldots \\
b_0 & a_1 & b_1 & 0 & 0 \ldots \\
0 & b_1 & a_2 & b_2 & 0 \ldots \\
\multicolumn{5}{c}{\cdots\cdots\cdots\cdots\cdots\cdots}
\end{array} \right\|
$$

is given. If we cross out the first p rows ($p = 1, 2, \ldots$) and the same number of columns we obtain the new matrix

$$
I_p = \left\| \begin{array}{cccccc}
a_p & b_p & 0 & 0 & 0 \ldots \\
b_p & a_{p+1} & b_{p+1} & 0 & 0 \ldots \\
0 & b_{p+1} & a_{p+2} & b_{p+2} & 0 \ldots \\
\multicolumn{5}{c}{\cdots\cdots\cdots\cdots\cdots\cdots}
\end{array} \right\|
$$

which is also a $\mathscr{J}$-matrix and may be called the p-th abbreviated matrix (in relation to the matrix I). The matrix I_p possesses two systems of polynomials namely, polynomials of the first kind $P_k^{(p)}(\lambda)$ and polynomials of the second kind $Q_k^{(p)}(\lambda)$; here $P_0^{(p)}(\lambda) = 1$ and $Q_0^{(p)}(\lambda) = 0$ for any p and $P_k^{(0)}(\lambda) = P_k(\lambda)$, $Q_k^{(0)}(\lambda) = Q_k(\lambda)$ for any k.

With the aid of the polynomials of the first kind one can construct the normalized positive sequence $\{s_k^{(p)}\}_k^\infty = 0$ belonging to a matrix I_p and therefore also the positive functional $S^{(p)}$ which acts in the space of all polynomials. With respect to this functional the polynomials of the first kind are orthogonal and normalized:

$$
S^{(p)}\{P_m^{(p)}(\lambda)P_n^{(p)}(\lambda)\} = \delta_{mn}.
$$

Prove that the orthogonal polynomials that belong to the p-th abbreviated matrix and the polynomials of the second kind that belong to the $(p-1)$-st abbreviated matrix are related by the equation

$$
P_k^{(p)}(\lambda) = b_{p-1}Q_{k+1}^{(p-1)}(\lambda) \quad (k = 0, 1, 2, \ldots).
$$

A consequence of this fact is that each of the abbreviated matrices is of the same type as the initial $\mathscr{J}$-matrix.

Chapter 2

THE POWER MOMENT PROBLEM

The present chapter is devoted to criteria of solubility and determinateness of the power moment problem and also to the investigation of certain analytic functions that occur in the indeterminate case. The method used is based on the quadrature formulae of Section 4 and the algebraic considerations of Section 2 of Chapter 1. In addition the last section of the chapter treats the moment problem in the light of functional analysis, as a problem of continuation of a positive functional.

1. Solubility criteria

1.1. We shall discuss Hamburger's moment problem. The statement of this problem is as follows: *Given an infinite sequence of real numbers $s_k (k = 0, 1, 2, ...) (s_0 = 1)$; it is required to find a non-decreasing function $\sigma(u) (-\infty < u < \infty)$ satisfying the equations*

$$s_k = \int_{-\infty}^{\infty} u^k \, d\sigma(u) \quad (k = 0, 1, 2, ...). \qquad ...[2.1]$$

This problem is not always soluble and therefore we shall first discuss the establishment of solubility criteria. In order not to have to deal with special algebraic considerations which become necessary when the integral in equation [2.1] degenerates into a finite sum, we shall introduce an additional requirement into our problem namely that the required function $\sigma(u)$ shall have an infinite number of points of increase.

We recall also that two solutions of the moment problem are not considered to be distinct if their difference is a constant at all points where that difference is continuous and also that the moment problem is called *determinate* if in this sense it has only a single solution, and *indeterminate* otherwise.

Our immediate problem is the proof of the following proposition:

29

THEOREM **2.1.1.** (HAMBURGER[3]) *The necessary and sufficient condition that must hold in order that a non-decreasing function* $\sigma(u)$ *($-\infty < u < \infty$), should exist having an infinite number of points of increase and such that*

$$\int_{-\infty}^{\infty} u^k \, d\sigma(u) = s_k \quad (k = 0, 1, 2, ...), \qquad ...[2.1]$$

is that the sequence $\{s_k\}_0^\infty$ *should be positive.*

It is very simple to prove that this condition is a necessary one. For if the equations [2.1] are satisfied, then for any m and any real $x_1, x_2, ..., x_m$ we have

$$\sum_{i,\,k=0}^{m} s_{i+k} x_i x_k = \int_{-\infty}^{\infty} \left[\sum_0^m x_k u^k \right]^2 d\sigma(u) \geqq 0. \qquad ...[2.2]$$

If the equality sign applies in this relation the polynomial

$$A(u) = \sum_0^m x_k u^k$$

must be zero at all points at which the function $\sigma(u)$ increases. But according to our condition there is an infinity of these points and therefore the polynomial $A(u)$ must be identically zero. Thus the equality sign is impossible in [2.2] if not all the $x_1, x_2, ..., x_m$ are zero. This completes the proof that the condition is necessary.

The proof that the condition is sufficient is more complicated and for convenience we shall split it in two parts.

a) We assume that the positive sequence $\{s_k\}_0^\infty$ is given. For any positive integer n we consider the so-called truncated moment problem of order $2n-1$, namely:

$$s_k = \int_{-\infty}^{\infty} u^k \, d\sigma(u) \quad (k = 0, 1, 2, ..., 2n-1), \qquad ...[2.3]$$

where $\sigma(u)$ is sought within the class of all non-decreasing functions. It is very simple to construct a whole continuum of solutions to this problem. Indeed, let us take an arbitrary real number τ ($-\infty < \tau < \infty$) and write down the equation proved in Section **4.2** of Chapter 1

$$\frac{Q_{n+1}(z) - \tau Q_n(z)}{P_{n+1}(z) - \tau P_n(z)} = \sum_1^{n+1} \frac{\mu_i}{z - \lambda_i}, \qquad ...[2.4]$$

where

$$\lambda_1 < \lambda_2 < ... < \lambda_{n+1} \quad [\lambda_i = \lambda_i^{(n+1)}(\tau)]$$

and

$$\mu_i = \mu_i^{(n+1)}(\tau) > 0.$$

As was established in the section just referred to we have

$$s_k = \sum_{i=1}^{n+1} \mu_i \lambda_i^k \quad (k = 0, 1, 2, ..., 2n-1). \qquad ...[2.5]$$

We now introduce the sectionally constant function $\sigma_n(u)$ which has as its only points of increase the λ_i, at which its discontinuities are the μ_i:

$$\mu_i = \sigma_n(\lambda_i + 0) - \sigma_n(\lambda_i - 0) \quad (i = 1, 2, ..., n+1);$$

then the equations [2.5] take on the form

$$s_k = \int_{-\infty}^{\infty} u^k \, d\sigma_n(u) \quad (k = 0, 1, 2, ..., 2n-1). \qquad ...[2.5a]$$

Thus the function $\sigma_n(u)$ we have constructed is a certain solution of the truncated moment problem [2.3].

We note that for $\tau < \infty$ the equations [2.5] and therefore the equations [2.5a] are also true for $k = 2n$, and if $\tau = 0$ also for $k = 2n+1$.

b) We now take any sequence of non-decreasing solutions $\sigma_n(u)$ of the truncated moment problem of order $2n-1$ $(n = 1, 2, 3, ...)$, for instance the sequence which was constructed above. This sequence $\{\sigma_n(u)\}_1^\infty$ satisfies the conditions of Helly's theorem† since for any n the total variation of the function $\sigma_n(u)$ is

$$\int_{-\infty}^{\infty} d\sigma_n(u) = s_0.$$

Therefore there exist such a non-decreasing function $\sigma(u)$ and such a sequence $\{\sigma_{n_i}(u)\}_{i=1}^\infty$ that at all points of continuity of $\sigma(u)$ the equation

$$\sigma(u) = \lim_{i \to \infty} \sigma_{n_i}(u)$$

holds. We prove that $\sigma(u)$ is a solution of the complete moment problem we are discussing, in other words that

$$s_k = \int_{-\infty}^{\infty} u^k \, d\sigma(u) \quad (k = 0, 1, 2, ...). \qquad ...[2.1]$$

Assume that $-A < 0$ and $B > 0$ are two points of continuity of $\sigma(u)$. In this case we have by Helly's second theorem that

$$\int_{-A}^{B} u^k \, d\sigma(u) = \lim_{i \to \infty} \int_{-A}^{B} u^k \, d\sigma_{n_i}(u). \qquad ...[2.6]$$

† For Helly's theorem see for instance LEVITAN[1] or GOKHMAN[1].

On the other hand if $2r$ is an even number greater than k then for $n_i > r$ we have

$$\int_{-\infty}^{\infty} u^k \, d\sigma_{n_i}(u) = \int_{-A}^{B} u^k \, d\sigma_{n_i}(u) + \int_{-\infty}^{-A} u^k \, d\sigma_{n_i}(u) + \int_{B}^{\infty} u^k \, d\sigma_{n_i}(u)$$

and

$$\left| \int_{-\infty}^{-A} + \int_{B}^{\infty} u^k \, d\sigma_{n_i}(u) \right| = \left| \int_{-\infty}^{-A} + \int_{B}^{\infty} \frac{u^{2r}}{u^{2r-k}} \, d\sigma_{n_i}(u) \right|$$

$$\leq \frac{1}{C^{2r-k}} \left\{ \int_{-\infty}^{-A} + \int_{B}^{\infty} u^{2r} \, d\sigma_{n_i}(u) \right\} \leq \frac{s_{2r}}{C^{2r-k}},$$

where $C = \min \{A, B\}$. Therefore it follows from [2.6] that

$$\left| \int_{-A}^{B} u^k \, d\sigma(u) - s_k \right| \leq \frac{s_{2r}}{C^{2r-k}}.$$

It now remains to increase A and B without limit, but in such a way that $-A$ and B always continue to be points of continuity of $\sigma(u)$. This proves equation [2.1].

The function $\sigma(u)$ which we have found evidently has an infinite number of points of increase. Indeed, if $\sigma(u)$ possessed only a finite number of points of increase (say $u_1, u_2, \ldots, u_N$), we would have for the polynomial

$$P(u) = (u-u_1)^2 (u-u_2)^2 \ldots (u-u_N)^2$$

the inequality

$$S\{P(u)\} = \int_{-\infty}^{\infty} (u-u_1)^2 (u-u_2)^2 \ldots (u-u_N)^2 \, d\sigma(u) = 0,$$

which is absurd because the given sequence $\{s_k\}_0^\infty$ is positive.

Thus the sufficiency of the condition of Theorem 2.1.1. is also proved.

1.2. We can amplify this result to some extent. To do so we recall the definition of the function $w_n(\lambda) = w_n(\lambda, \tau)$ introduced in Section 2.3 of Chapter 1. By virtue of that definition the left hand side of [2.4] is $-w_{n+1}(z)$. Therefore, using the functions $\sigma_n(u)$ constructed in Section 1.1, equation [2.4] can be written in the form

$$w_{n+1}(z) = \int_{-\infty}^{\infty} \frac{d\sigma_n(u)}{u-z} \quad (\text{Im } z \neq 0).$$

We saw above that the point $w_{n+1} = w_{n+1}(z)$ describes the circumference of the circle $K_{n+1}(z)$ when the parameter τ varies over the real axis. The following fact emerges from our construction: let us fix a non-real point z; then for any point w_{n+1} lying on the circumference $K_{n+1}(z)$ there exists a solution $\sigma_n(u)$ of the truncated moment problem [2.3] such that

$$\int_{-\infty}^{\infty} \frac{d\sigma_n(u)}{u-z} = w_{n+1}.$$

Indeed in order to obtain the required solution $\sigma_n(u)$ one must merely determine the real parameter τ in such a way that

$$w_{n+1} = -\frac{Q_{n+1}(z)-\tau Q_n(z)}{P_{n+1}(z)-\tau P_n(z)}.$$

We take a fixed non-real value of z, and for $n = 1, 2, 3, \ldots$ we take points w_n on the circles $K_n(z)$. If the case of a limiting circle prevails we take these points w_n in such a way that the sequence $\{w_n\}_1^\infty$ converges to a pre-determined point of the limiting circumference $K_\infty(z)$. In this way we get the equations

$$s_k = \int_{-\infty}^{\infty} u^k\, d\sigma_n(u) \quad (k = 0, 1, \ldots, 2n-1),$$

$$\lim_{n \to \infty} \int_{-\infty}^{\infty} \frac{d\sigma_n(u)}{u-z} = w. \qquad \ldots[2.7]$$

We shall apply the construction given in (b) of Section 1.1 to the sequence $\sigma_n(u)$, which satisfies [2.7]. Helly's second theorem is also applicable to this equation. Indeed,

$$\int_{-\infty}^{\infty} \frac{d\sigma_{n_i}(u)}{u-z} = \int_{-A}^{B} \frac{d\sigma_{n_i}(u)}{u-z} + \varepsilon_{A,\,B},$$

where

$$|\varepsilon_{A,\,B}| \leq \int_{-\infty}^{-A} + \int_{B}^{\infty} \frac{d\sigma_{n_i}(u)}{|u-z|} \leq \frac{1}{C}\int_{-\infty}^{\infty}\left|\frac{u}{u-z}\right|\,d\sigma_{n_i}(u) \leq \frac{L}{C}s_0,$$

and the quantity

$$L = \sup_{-\infty < u < \infty}\left|\frac{u}{u-z}\right|$$

depends only on z.

Thus we have incidentally proved the following proposition:

Assume that the positive sequence $\{s_k\}_0^\infty$ has a $\mathscr{J}$-matrix of type C corresponding to it; in this case one can construct a solution to the moment problem

$$s_k = \int_{-\infty}^{\infty} u^k \, d\sigma(u) \quad (k = 0, 1, 2, \ldots), \qquad \ldots [2.1]$$

for which the quantity

$$\int_{-\infty}^{\infty} \frac{d\sigma(u)}{u-z}$$

will fall on a pre-determined point w of the circumference $K_\infty(z)$.

The following theorem is therefore true:

THEOREM 2.1.2. *If the positive sequence $\{s_k\}_0^\infty$ has a $\mathscr{J}$-matrix of type C corresponding to it, then the moment problem $[2.1]$ is indeterminate.*

The inverse proposition is also true; we shall prove it in the following section.

2. The isometric operator generated by orthogonal polynomials

2.1. Assume we have a certain positive sequence $\{s_k\}_0^\infty$ and that some solution $\sigma(u)$ of the moment problem

$$s_k = \int_{-\infty}^{\infty} u^k \, d\sigma(u) \quad (k = 0, 1, 2, \ldots) \qquad \ldots [2.8]$$

is known. We shall construct the space L_σ^p ($p \geq 1$) of all σ-measurable functions $f(u)$ for which

$$\int_{-\infty}^{\infty} |f(u)|^p \, d\sigma(u) < \infty.$$

An important property of the space L_σ^p is its completeness. It has the significance that for any sequence of functions $f_n(u) \in L_\sigma^p$ satisfying the relation

$$\lim_{m, n \to \infty} \int_{-\infty}^{\infty} |f_n(u) - f_m(u)|^p \, d\sigma(u) = 0$$

(i.e. convergent in the Cauchy sense) there exists a function $f(u) \in L_\sigma^p$ for which

$$\lim_{n \to \infty} \int_{-\infty}^{\infty} |f(u) - f_n(u)|^p \, d\sigma(u) = 0,$$

i.e. a function $f(u) \in L_\sigma^p$ to which the given sequence converges.

The completeness proof for a space L_σ^p in which the norm is of the form

$$\left\{ \int_{-\infty}^{\infty} |f(u)|^p \, du \right\}^{\frac{1}{p}},$$

is given in many books.†

The completeness proof for the space L_σ^p can be carried through by the same method.

In the first place we shall be interested in the space L_σ^p which is a Hilbert space, if the scalar product is defined by the formula

$$(f, g)_\sigma = \int_{-\infty}^{\infty} f(u)\overline{g(u)} \, d\sigma(u).$$

The orthogonal polynomials $P_k(u)$ form an orthonormal system in L_σ^2.

Together with L_σ^2 we consider a Hilbert space l^2. Its elements are numerical (complex) sequences

$$x = \{x_0, x_1, x_2, ...\}, \quad y = \{y_0, y_1, y_2, ...\}, \quad ... ,$$

for which

$$\sum_0^\infty |x_k|^2 < \infty, \quad \sum_0^\infty |y_k|^2 < \infty, \quad ... ,$$

The scalar product in this space is defined by

$$(x, y) = \sum_0^\infty x_k \bar{y}_k.$$

We shall call a vector $x \in l^2$ *finite* if it has only a finite number of components different from zero.

2.2. We shall discuss a certain mapping of the space l^2 on L_σ^2. We shall denote by U the operator effecting this mapping. It will be constructed in two steps. In the first place we shall define the operator U over all finite vectors x by the formula

$$f(u) = Ux = x_0 P_0(u) + x_1 P_1(u) + \ ... \ + x_n P_n(u) + \ ...$$

Since this series terminates, so that $f(u)$ is a polynomial, we may write down the following equality:

$$\int_{-\infty}^{\infty} |f(u)|^2 \, d\sigma(u) = \sum_{i,k=0}^{\infty} x_i \cdot \bar{x}_k \int_{-\infty}^{\infty} P_i(u) P_k(u) \, d\sigma(u)$$

$$= \sum_0^\infty |x_k|^2, \qquad ...[2.9]$$

† See for instance AKHIEZER[3].

in which the series also terminate. This equality can be stated in the form

$$(f, f)_\sigma = (x, x).$$

Similarly, for two finite vectors x and y and two functions (polynomials) $f(u) = Ux$ and $g(u) = Uy$ we shall have the equation

$$(f, g)_\sigma = (x, y).$$

We now go over to arbitrary, no longer finite, vectors of l^2. If x is such a vector we can, for any n, introduce the function

$$f_n(u) = x_0 P_0(u) + x_1 P_1(u) + \ldots + x_n P_n(u).$$

Applying the relation $[2.9]$ to the difference

$$f_n(u) - f_m(u) = x_{m+1} P_{m+1}(u) + \ldots + x_n P_n(u),$$

we obtain

$$\int_{-\infty}^{\infty} |f_n(u) - f_m(u)|^2 \, d\sigma(u) = |x_{m+1}|^2 + |x_{m+2}|^2 + \ldots + |x_n|^2.$$

Hence it follows that

$$\lim_{m, n \to \infty} \int_{-\infty}^{\infty} |f_n(u) - f_m(u)|^2 \, d\sigma(u) = 0,$$

i.e. the sequence of functions $f_n(u) \in L_\sigma^2$ converges in the Cauchy sense. By virtue of the completeness of L_σ^2 there will exist an element $f(u) \in L_\sigma^2$, which is evidently unique, to which the sequence $\{f_n(u)\}_1^\infty$ converges. This element $f(u)$ may be written in the form

$$f(u) = \underset{\sigma(u)}{\text{l.i.m.}} f_n(u).$$

As a matter of definition we shall take $f(u)$ to be Ux. Thus we have for any $x \in l^2$

$$f(u) = Ux = \underset{\sigma(u)}{\text{l.i.m.}} \sum_{k}^{n} x_k P_k(u).$$

We have defined the operator U everywhere in l^2 but have left open the question whether the linear manifold Δ_U on to which the operator U maps l^2 coincides with L_σ^2. This question will be solved in the following sections.

However we can assert that the operator U is *isometric*, i.e. for any two vectors $x, y \in l^2$ the equality

$$(f, g)_\sigma = (x, y)$$

holds, where

$$f(u) = Ux, \quad g(u) = Uy.$$

Indeed, this follows directly from the fact that

$$(x, y) = \lim_{n \to \infty} \sum_0^n x_k \bar{y}_k = \lim_{n \to \infty} (f_n, g_n)_\sigma,$$

since

$$(f_n, g_n)_\sigma = (f, g)_\sigma + (f_n - f, g)_\sigma + (f, g_n - g)_\sigma + (f_n - f, g_n - g)_\sigma$$

and therefore

$$|(f_n, g_n)_\sigma - (f, g)_\sigma|$$

$$\leqq \| f_n - f \|_\sigma \| g \|_\sigma + \| f \|_\sigma \| g_n - g \|_\sigma + \| f_n - f \|_\sigma \| g_n - g \|_\sigma,$$

and the right hand side tends to zero for $n \to \infty$.

We shall formulate our result in the form of the following proposition.

THEOREM 2.2.2. *Any solution $\sigma(u)$ of the moment problem* (1) *generates an operator U according to the formula*

$$f(u) = Ux = \underset{\sigma(u)}{\text{l.i.m.}} \sum_0^n x_k P_k(u). \qquad \qquad \ldots[2.10]$$

The domain of definition of this operator is the whole space l^2 and the whole range of its values a certain manifold $\Delta_U \subset L_\sigma^2$. This operator is isometric:

$$(Ux, Uy)_\sigma = \int_{-\infty}^\infty f(u)\overline{g(u)} \, d\sigma(u) = \sum_0^\infty x_k \bar{y}_k = (x, y).$$

The inversion of equation [2.10] has the form

$$x_k = \int_{-\infty}^\infty f(u)P_k(u) \, d\sigma(u) \quad (k = 0, 1, 2, \ldots). \qquad \ldots[2.11]$$

To prove the theorem we only have to show that the last statement holds. But it follows from the fact that for any $n > k$

$$x_k = \int_{-\infty}^\infty f_n(u)P_k(u) \, d\sigma(u) = (f_n, P_k)_\sigma$$

and

$$\lim_{n \to \infty} (f_n, P_k)_\sigma = (f, P_k)_\sigma.$$

2.3. The expressions [2.11] can be called the generalized Fourier coefficients of $f(u)$. They can be introduced for any function $F(u) \in L^2_\sigma$ and not only for functions $f(u) \in \Delta_U$, as was done in Section **2.2**. Thus, with any function $F(u) \in L^2_\sigma$ one can associate a generalized Fourier series

$$F(u) \sim \sum_0^\infty x_k P_k(u) \quad [x_k = (F, P_k)_\sigma].\qquad\qquad ...[2.12]$$

With the aid of simple arguments that are well known from the theory of ordinary Fourer series and which are repeated in a general form in operator theory, the following extremal property for partial sums of the series [2.12] can be established: Among all polynomials $R_n(u)$ of degree $\leqq n$ the partial sum

$$\sum_0^n x_k P_k(u)$$

is the best approximation to the function $F(u)$ in the metric of L^2_σ, i.e

$$I_n \equiv \int_{-\infty}^\infty \left| F(u) - \sum_0^n x_k P_k(u) \right|^2 d\sigma(u)$$

$$= \min_{R_n} \int_{-\infty}^\infty | F(u) - R_n(u) |^2 d\sigma(u).\qquad\qquad ...[2.13]$$

Further it can be established by simple calculation that

$$I_n = \int_{-\infty}^\infty | F(u) |^2 d\sigma(u) - \sum_0^n | x_k |^2,\qquad\qquad ...[2.14]$$

whence follows the inequality

$$\sum_0^\infty | x_k |^2 \leqq \int_{-\infty}^\infty | F(u) |^2 d\sigma(u),$$

which may be called the Bessel inequality. If the equality sign is valid in this relation it is called Parseval's equality. From a comparison of equations [2.13] and [2.14] it follows that the validity of Parseval's equality for some function $F(u) \in L^2_\sigma$ signifies that this function can be approximated in L^2_σ with any degree of accuracy, by means of a polynomial.

THEOREM **2.2.3.** *The linear manifold* Δ_U *coincides with* L^2_σ *if and only if the set of all polynomials is dense in* L^2_σ.

Proof. Assume that the set of all polynomials is dense in L^2_σ. We take an arbitrary function $F(u) \in L^2_\sigma$ and assume that

$$F(u) \sim \sum_0^\infty x_k P_k(u).$$

We prove that the function

$$f(u) = Ux,$$

where $x = \{x_0, x_1, x_2, ...\}$, coincides with $F(u)$, whence it will follow that $F(u) \in \Delta_U$. Let us assume that the converse holds, i.e. that

$$\int_{-\infty}^\infty |F(u) - f(u)|^2 \, d\sigma(u) \neq 0.$$

If we write down the Parseval equality for the function $F(u) - f(u)$ the result is absurd, since

$$x_k = (F, P_k)_\sigma = (f, P_k)_\sigma$$

and therefore

$$(F - f, P_k) = 0 \quad (k = 0, 1, 2, ...).$$

Thus the sufficiency of the condition is proved.

We now prove its necessity. Assume that it is known that $\Delta_U \in L^2_\sigma$. In this case every function $F(u) \in L^2_\sigma$ can be written in the form

$$F(u) = Ux$$

and in L^2_σ it is therefore the limit of a sequence of polynomials

$$F_n(u) = \sum_0^n x_k P_k(u).$$

This proves that the condition is also necessary.

If the range of values of the isometric operator is the whole of space, the operator is called *unitary*.

Thus the following is true:

COROLLARY 2.2.3. *The operator U is unitary if and only if the Parseval equality*

$$\int_{-\infty}^\infty |F(u)|^2 \, d\sigma(u) = \sum_0^\infty |x_k|^2 \quad [x_k = (F, P_k)_\sigma]$$

holds for any function $F(u) \in L^2_\sigma$.

2.4. We shall utilize these results for a preliminary investigation of the set of all solutions of the moment problem [2.8]. Together with each of these solutions we shall consider the functions

$$w(\lambda) = \int_{-\infty}^{\infty} \frac{d\sigma(u)}{u-\lambda} \quad (\text{Im } \lambda \neq 0).$$

THEOREM 2.2.4.† *The set of values assumed at the point λ (Im $\lambda \neq 0$) by all functions $w(\lambda)$ belonging to a moment problem, coincides with the point set $K_\infty(\lambda)$.*

Proof. Assume that w is the value of the function $w(\lambda)$ corresponding to some solution $\sigma(u)$ of the moment problem,

We find the generalized Fourier series of the function

$$f(u) = \frac{1}{u-\lambda}.$$

Since

$$\int_{-\infty}^{\infty} \frac{1}{u-\lambda} P_k(u)\, d\sigma(u) = \int_{-\infty}^{\infty} \frac{P_k(u)-P_k(\lambda)}{u-\lambda}\, d\sigma(u) + P_k(\lambda)\int_{-\infty}^{\infty} \frac{d\sigma(u)}{u-\lambda}$$

$$= Q_k(\lambda) + wP_k(\lambda),$$

we have

$$\frac{1}{u-\lambda} \sim \sum_0^{\infty} \{wP_k(\lambda) + Q_k(\lambda)\}P_k(u)$$

and Bessel's inequality can be written in the form

$$\sum_0^{\infty} |wP_k(\lambda) + Q_k(\lambda)|^2 \leq \int_{-\infty}^{\infty} \frac{d\sigma(u)}{|u-\lambda|^2}. \qquad \ldots [2.15]$$

But

$$\int_{-\infty}^{\infty} \frac{d\sigma(u)}{|u-\lambda|^2} = \int_{-\infty}^{\infty} \frac{1}{\lambda-\bar{\lambda}}\left\{\frac{1}{u-\lambda} - \frac{1}{u-\bar{\lambda}}\right\} d\sigma(u) = \frac{w-\bar{w}}{\lambda-\bar{\lambda}}.$$

Therefore the inequality [2.15] expresses the fact that $w \in K_\infty(\lambda)$.

It must now be proved that if $w \in K_\infty(\lambda)$ then a solution $\sigma(u)$ of the moment problem can be found for which

$$\int_{-\infty}^{\infty} \frac{d\sigma(u)}{u-\lambda} = w.$$

For the case when $K_\infty(\lambda)$ is a point and also for the case when $K_\infty(\lambda)$ is a circle and when the point w lies on the boundary of this circle this

† NEVANLINNA[1], M. RIESZ[3].

fact has already been proved in the previous section. Therefore consider the case that $K_\infty(\lambda)$ is a circle and the point w lies within the circle. We draw a chord through the point w and find two points w_1 and w_2 on the boundary of the circle $K_\infty(\lambda)$. In terms of these points the point w may be expressed as

$$w = \theta w_1 + (1 - \theta) w_2,$$

where

$$0 < \theta < 1.$$

We can find solutions $\sigma_1(u)$ and $\sigma_2(u)$ for which

$$\int_{-\infty}^{\infty} \frac{d\sigma_1(u)}{u - \lambda} = w_1, \quad \int_{-\infty}^{\infty} \frac{d\sigma_2(u)}{u - \lambda} = w_2.$$

We then introduce the function

$$\sigma(u) = \theta \sigma_1(u) + (1 - \theta)\sigma_2(u),$$

which evidently is a certain solution of our moment problem. Also, the equality

$$\int_{-\infty}^{\infty} \frac{d\sigma(u)}{u - \lambda} = \theta w_1 + (1 - \theta) w_2 = w$$

will hold, which just provides the proof of our statement.

COROLLARY 2.2.4.† *If the $\mathscr{J}$-matrix corresponding to the sequence $\{s_k\}_0^\infty$ is of type D the moment problem*

$$s_k = \int_{-\infty}^{\infty} u^k \, d\sigma(u) \quad (k = 0, 1, 2, \ldots)$$

is determinate.

Proof. We take two solutions of the moment problem, $\sigma_1(u)$ and $\sigma_2(u)$ and consider the functions

$$w_1(\lambda) = \int_{-\infty}^{\infty} \frac{d\sigma_1(u)}{u - \lambda}, \quad w_2(\lambda) = \int_{-\infty}^{\infty} \frac{d\sigma_2(u)}{u - \lambda} \quad (\text{Im } \lambda \neq 0).$$

These functions must coincide for any λ (Im $\lambda \neq 0$), since the set $K_\infty(\lambda)$ which contains the points $w_1(\lambda)$ and $w_2(\lambda)$ consists by assumption of a single point for any such λ. Putting

$$\omega(u) = \sigma_1(u) - \sigma_2(u),$$

† This is the converse of Theorem 2.1.2.

we find that

$$\int_{-\infty}^{\infty} \frac{d\omega(u)}{u-\lambda} = 0 \quad (\operatorname{Im} \lambda \neq 0).$$

Using the inversion formula of Stieltjes and Perron we† find that

$$\frac{\omega(u-0)+\omega(u+0)}{2} = \text{const} \quad (-\infty < u < \infty)$$

and therefore $\sigma_1(u)$ and $\sigma_2(u)$ represent one and the same solution of our moment problem.

3. Some criteria of completeness

3.1. Assume that a non-decreasing function $\sigma(u)$ $(-\infty < u < \infty)$ is given for which the integrals

$$\int_{-\infty}^{\infty} u^k \, d\sigma(u) \quad (k = 0, 1, 2, \ldots)$$

converge. In this case any one of the spaces $L_\sigma^p (p \geqq 1)$ generated by the functions $\sigma(u)$ contains the set of all polynomials and, naturally, the question arises as to the conditions under which this set is dense in L_σ^p. Since the requirement imposed on the function $\sigma(u)$ is such that from it a certain Hamburger moment problem can be constructed, of which $\sigma(u)$ is the solution (or one of the solutions), one can attempt to solve the problem stated in terms of the moment problem. The present section is devoted to such a solution for $p = 2$ and $p = 1$. Among the tools that we shall use we mention the general Banach criterion which amounts to the following:‡ In order that a linear set $\mathfrak{M}$ be dense in a linear normed space E it is necessary and sufficient that any linear functional in E which is zero on every element of the set $\mathfrak{M}$ be zero identically.

3.2. Assume that $\sigma(u)$ is a solution of the power moment problem. We take any non-real λ and consider the point

$$w = \int_{-\infty}^{\infty} \frac{d\sigma(u)}{u-\lambda}. \qquad \ldots[2.16]$$

As was established above, the inequality

$$\sum_0^\infty | wP_k(\lambda)+Q_k(\lambda) |^2 \leqq \frac{w-\bar{w}}{\lambda-\bar\lambda} \qquad \ldots[2.17]$$

---

† This formula and also its proof are given at the end of Chapter 3 (see Addenda and Problems to Chapter 3, no. 1).
‡ See BANACH[1] and also AKHIEZER[3].

SOME CRITERIA OF COMPLETENESS

holds. If the equality sign is valid in this relation we shall call the solution $\sigma(u)$ *N-extremal at the point* λ. We shall see immediately that the words " at the point λ " can be omitted, in other words that the *N*-extremal property of a solution at a single point λ (Im $\lambda \neq 0$) always implies the *N*-extremal property at any point of this class.

THEOREM 2.3.2. (M. RIESZ[4]). *In order that the set of all polynomials in L_σ^2 be dense, it is necessary that $\sigma(u)$ as a solution of the moment problem be N-extremal at every non-real point and it is sufficient that it should be N-extremal in at least one such point.*

The proof of the necessary condition is very simple. Assume that for some non-real value of λ the inequality holds:

$$\sum_0^\infty |\, wP_k(\lambda) + Q_k(\lambda) \,|^2 < \frac{w - \overline{w}}{\lambda - \overline{\lambda}}. \qquad \ldots[2.18]$$

Since the

$$wP_k(\lambda) + Q_k(\lambda) \quad (k = 0, 1, 2, \ldots)$$

are the coefficients of the Fourier series in the polynomials $P_k(u)$ of the function

$$f_0(u) = \frac{1}{u - \lambda}$$

and

$$\frac{w - \overline{w}}{\lambda - \overline{\lambda}} = \int_{-\infty}^\infty \frac{1}{|\, u - \lambda \,|^2}\, d\sigma(u) = \int_{-\infty}^\infty |f_0(u)|^2\, d\sigma(u),$$

the inequality $[2.18]$ signifies that the Parseval equality is not valid for the function $f_0(u)$. Therefore $f_0(u)$ cannot be approximated to any degree of accuracy in terms of polynomials in L_σ^2, which means that the set of all polynomials is not dense in L_σ^2.

Passing to the proof of sufficiency, we assume that the solution $\sigma(u)$ of the moment problem is *N*-extremal at a certain non-real point λ, so that

$$\frac{w - \overline{w}}{\lambda - \overline{\lambda}} = \sum_0^\infty |\, wP_k(\lambda) + Q_k(\lambda) \,|^2.$$

Hence it follows that the function $f_0(u)$ can be approximated in L_σ^2 by a polynomial, to any degree of accuracy. The same is evidently true for the function

$$\bar{f}_0(u) = \frac{1}{u - \overline{\lambda}}.$$

Further

$$\int_{-\infty}^{\infty} \left| \frac{1}{(u-\lambda)^2} - \frac{A}{u-\lambda} - \sum_{0}^{n} B_k u^k \right|^2 d\sigma(u)$$

$$\leq \frac{1}{\eta^2} \int_{-\infty}^{\infty} \left| \frac{1}{u-\lambda} - A - (u-\lambda) \sum_{0}^{n} B_k u^k \right|^2 d\sigma(u),$$

where

$$\eta = \operatorname{Im} \lambda,$$

and it has been established that the right hand side can be made arbitrarily small. Therefore the functions

$$f_1(u) = \frac{1}{(u-\lambda)^2}, \quad \bar{f}_1(u) = \frac{1}{(u-\bar{\lambda})^2}$$

can also be approximated to any accuracy by polynomials. Repeating the same arguments we find that the result obtained is also true for all subsequent powers

$$f_k(u) = \frac{1}{(u-\lambda)^{k+1}}, \quad \bar{f}_k(u) = \frac{1}{(u-\bar{\lambda})^{k+1}} \quad (k = 2, 3, \ldots).$$

We wish to prove that the set of all polynomials is dense in L_σ^2. Assuming the opposite, we find a certain linear functional $\Phi(f)$ in L_σ^2 which becomes zero for all polynomials but is not identically zero. By a well known theorem of functional analysis the functional $\Phi(f)$ has the form

$$\Phi(f) = \int_{-\infty}^{\infty} f(u)g(u) \, d\sigma(u),$$

where $f(u)$ is an arbitrary element of the space L_σ^2 and the function $g(u) \in L_\sigma^2$ generates the functional $\Phi(f)$. Since $\Phi(f)$ is not identically zero we have

$$\int_{-\infty}^{\infty} |g(u)|^2 \, d\sigma(u) \neq 0, \qquad \qquad \ldots [2.19]$$

and on the other hand, according to our assumption we have

$$\int_{-\infty}^{\infty} g(u)u^k \, d\sigma(u) = 0 \quad (k = 0, 1, 2, \ldots).$$

Using these equations and the fact that the functions

$$f_k(u), \quad \bar{f}_k(u) \quad (k = 0, 1, 2, \ldots)$$

can be arbitrarily closely approximated by polynomials in L_σ^2 it is easy
to conclude that

$$\int_{-\infty}^{\infty} \frac{g(u)}{(u-\lambda)^k} \, d\sigma(u) = 0,$$

$$(k = 1, 2, 3, \ldots).$$

$$\int_{-\infty}^{\infty} \frac{\overline{g(u)}}{(u-\lambda)^k} \, d\sigma(u) = 0$$

We introduce the functions

$$\phi(z) = \int_{-\infty}^{\infty} \frac{g(u)}{u-z} \, d\sigma(u), \quad \psi(z) = \int_{-\infty}^{\infty} \frac{\overline{g(u)}}{u-z} \, d\sigma(u).$$

They are analytic in the half space Im $z > 0$ and by what has been proved
become zero together with all their derivatives at the point $z = \lambda$.
Therefore $\phi(z) \equiv 0$ and $\psi(z) \equiv 0$. By virtue of the Stieltjes-Perron
inversion formula we have therefore that

$$\int_{-\infty}^{u} g(t) \, d\sigma(t) = 0 \quad (-\infty < u < \infty),$$

and this contradicts [2.19]. Thus we have also proved the sufficiency
condition.

3.3. The result of the previous section can be given a more convenient formulation. To do this we make the following definition:

DEFINITION 2.3.3. The solution $\sigma(u)$ of the power moment problem
is called *N-extremal* if one of the two following conditions holds:

a) This solution is unique,

b) The solution is not unique but the point

$$w = \int_{-\infty}^{\infty} \frac{d\sigma(u)}{u-\lambda} \qquad \ldots[2.16]$$

lies on the circumference of the circle $K_\infty(\lambda)$ for some (and therefore for
any) non-real λ.

The new formulation of Theorem 2.3.2. is as follows:

THEOREM 2.3.3. *The set of all polynomials is dense in L_σ^2 if and only if
$\sigma(u)$ is an N-extremal solution of the moment problem.*

A consequence of this theorem is

COROLLARY 2.3.3. *If $\sigma(u)$ is the solution of a determinate moment
problem the set of all polynomials is dense in L_σ^2.*

3.4. The aggregate V of all solutions of the moment problem is a certain convex set in the space of all non-decreasing functions. This means that if

$$\sigma_1(u) \in V, \quad \sigma_2(u) \in V,$$

then for any constant $\alpha_1 > 0$ and $\alpha_2 > 0$ we have

$$\sigma(u) = \frac{\alpha_1 \sigma_1(u) + \alpha_2 \sigma_2(u)}{\alpha_1 + \alpha_2} \in V.$$

It is natural to consider the limit points of the set V. As we shall see, this is necessary in order to obtain a criterion for the aggregate of all polynomials in L_σ^2 to be dense.

DEFINITION **2.3.4.** A solution of a moment problem $\sigma(u)$ is called *V-extremal* if it cannot be represented in the form

$$\sigma(u) = \frac{\alpha_1 \sigma_1(u) + \alpha_2 \sigma_2(u)}{\alpha_1 + \alpha_2},$$

where $\alpha_1 > 0, \alpha_2 > 0$ and $\sigma_1(u)$ and $\sigma_2(u)$ are solutions of the same moment problem, different from $\sigma(u)$.

It is not difficult to prove that any *N*-extremal solution of the moment problem is *V*-extremal.†

Indeed, assume that $\sigma(u)$ is not a *V*-extremal solution and that therefore it permits representation in the form given in Definition **2.3.4.** In this case the functions

$$w(\lambda) = \int_{-\infty}^{\infty} \frac{d\sigma(u)}{u - \lambda}, \quad w_1(\lambda) = \int_{-\infty}^{\infty} \frac{d\sigma_1(u)}{u - \lambda}, \quad w_2(\lambda) = \int_{-\infty}^{\infty} \frac{d\sigma_2(u)}{u - \lambda},$$

where Im $\lambda \neq 0$, are connected by the relation

$$w(\lambda) = \frac{\alpha_1 w_1(\lambda) + \alpha_2 w_2(\lambda)}{\alpha_1 + \alpha_2}.$$

Since $w_1(\lambda) \not\equiv w_2(\lambda)$ there must be a value of λ for which the points $w_1(\lambda)$ and $w_2(\lambda)$, which lie in the closed circle $K_\infty(\lambda)$, are different. But in such a case the point $w(\lambda)$ cannot lie on the boundary of the circle $K_\infty(\lambda)$, by virtue of the inequalities $\alpha_1 > 0$ and $\alpha_2 > 0$. Therefore the solution $\sigma(u)$ is not *N*-extremal.

Is it possible to characterize a *V*-extremal solution by a property analogous to that which according to Theorem **2.3.3.** characterizes an *N*-extremal solution?

An answer to this question is given by the following theorem.

† It will be shown below that the inverse statement is untrue.

THEOREM **2.3.4.** (NAIMARK[3]). *The set of all polynomials is dense in L_σ^2 if and only if $\sigma(u)$ is a V-extremal solution of a moment problem.*

Proof.† Assume that $\sigma(u)$ is not a V-extremal solution of a moment problem and therefore that it allows the representation

$$\sigma(u) = \alpha\sigma_1(u) + (1-\alpha)\sigma_2(u) \quad (0 < \alpha < 1),$$

where α is a constant and $\sigma_1(u)$ and $\sigma_2(u)$ are a pair of different solutions of the same moment problem. Denoting by $\|f\|$ the norm in L_σ^1 and putting

$$\Phi(f) = \int_{-\infty}^{\infty} f(u)\,d\sigma(u), \quad \Phi_1(f) = \int_{-\infty}^{\infty} f(u)\,d\sigma_1(u), \quad \ldots[2.20]$$

where $f(u) \in L_\sigma^1$, we find that

$$\Phi(f) \leq \int_{-\infty}^{\infty} |f(u)|\,d\sigma(u) = \|f\|$$

and

$$|\Phi_1(f)| \leq \int_{-\infty}^{\infty} |f(u)|\,d\sigma_1(u) \leq \frac{1}{\alpha}\int_{-\infty}^{\infty} |f(u)|\,d\sigma(u) = \frac{1}{\alpha}\|f\|.$$

Therefore the integrals [2.20] are linear functionals in L_σ^1. But then the following will also be a linear functional:

$$\Phi_0(f) = \Phi(f) - \Phi_1(f).$$

This functional is evidently not identically zero. At the same time it becomes zero for all polynomials. Hence it follows by virtue of the theorem in functional analysis mentioned in Section **3.1** that the set of all polynomials is not dense in L_σ^1.

Conversely assume it to be given that the set of all polynomials is not dense in L_σ^1. It then follows from the same theorem of functional analysis that there exists in L_σ^1 a functional, say $\Phi_0(f)$, which is not identically zero but which is zero for all polynomials. We can assume that the norm of this functional is 1. We put

$$\left.\begin{aligned}\Phi_1(f) &= \int_{-\infty}^{\infty} f(u)\,d\sigma(u) - \Phi_0(f),\\ \Phi_2(f) &= \int_{-\infty}^{\infty} f(u)\,d\sigma(u) + \Phi_0(f).\end{aligned}\right\} \quad \ldots[2.21]$$

† The proof given here is due to Gel'fand (see NAIMARK[3]).

These functionals are positive (more precisely, non-negative). Indeed, if at any point at which the function $\sigma(u)$ increases the inequality $f(u) \geqq 0$ holds, then

$$\int_{-\infty}^{\infty} f(u)\, d\sigma(u) = \int_{-\infty}^{\infty} |f(u)|\, d\sigma(u) = \| f \|,$$

while

$$| \Phi_0(f) | \leqq \| f \|,$$

because the norm of the functional $\Phi_0(f)$ is equal to 1 by assumption. Bearing in mind the form of a general linear positive functional in L_σ^1 we conclude that the functionals $\Phi_1(f)$ and $\Phi_2(f)$ permit the representation

$$\Phi_1(f) = \int_{-\infty}^{\infty} f(u)g_1(u)\, d\sigma(u), \quad \Phi_2(f) = \int_{-\infty}^{\infty} f(u)g_2(u)\, d\sigma(u),$$

where the functions $g_1(u)$ and $g_2(u)$ are real, non-negative, σ-measurable and bounded. In other words

$$\Phi_1(f) = \int_{-\infty}^{\infty} f(u)\, d\sigma_1(u), \quad \Phi_2(f) = \int_{-\infty}^{\infty} f(u)\, d\sigma_2(u),$$

where

$$\sigma_1(u) = \int_{-\infty}^{u} g_1(t)\, d\sigma(t), \quad \sigma_2(u) = \int_{-\infty}^{u} g_2(t)\, d\sigma(t)$$

are two non-decreasing functions of bounded variation. By the equations [2.21] we have

$$\left. \begin{aligned} \int_{-\infty}^{\infty} f(u)\, d\sigma(u) &= \int_{-\infty}^{\infty} f(u)\, d\sigma_1(u) + \Phi_0(f), \\ \int_{-\infty}^{\infty} f(u)\, d\sigma(u) &= \int_{-\infty}^{\infty} f(u)\, d\sigma_2(u) - \Phi_0(f). \end{aligned} \right\} \quad \dots[2.22]$$

Since the functional $\Phi_0(f)$ becomes zero on any polynomial $f(u)$, we have

$$\int_{-\infty}^{\infty} u^k\, d\sigma(u) = \int_{-\infty}^{\infty} u^k\, d\sigma_1(u) = \int_{-\infty}^{\infty} u^k\, d\sigma_2(u) \quad (k = 0, 1, 2, \dots).$$

Therefore $\sigma_1(u)$ and $\sigma_2(u)$ are solutions of our moment problem and are different because $\Phi_1(f) \not\equiv 0$. But it follows from [2.22] that for any function $f(u) \in L_\sigma^1$ we have

$$\int_{-\infty}^{\infty} f(u)\, d\sigma(u) = \frac{1}{2} \int_{-\infty}^{\infty} f(u)\, d\sigma_1(u) + \frac{1}{2} \int_{-\infty}^{\infty} f(u)\, d\sigma_2(u).$$

Therefore

$$\sigma(u) = \tfrac{1}{2}\sigma_1(u) + \tfrac{1}{2}\sigma_2(u)$$

and thus the solution $\sigma(u)$ is not V-extremal.

Thus the theorem is proved.

4. The function $\rho(z)$ and the Nevanlinna matrices

4.1. For any real or non-real z we put

$$\rho(z) = \frac{1}{\displaystyle\sum_0^{\infty} | P_k(z) |^2}$$

and

$$\rho_n(z) = \frac{1}{\displaystyle\sum_0^{n} | P_k(z) |^2}$$

as was shown in Section **2.3** of Chapter 1 the radius of the circle $K_{n+1}(z)$ (Im $z \neq 0$) is

$$\frac{1}{| z - \bar{z} |}\, \rho_n(z)$$

and similarly $\rho(z)$ determines the radius of the circle $K_\infty(z)$.

Further, the function $\rho_n(z)$ occurred in Section **4.1** of Chapter 1 in the construction of the quadrature formulae. Indeed, the quadrature formula [1.32] of Section **4**, Chapter 1 can be written in the form

$$S\{R(\lambda)\} = \sum_{k=1}^{n} \rho_{n-1}(\lambda_k) R(\lambda_k).$$

We shall now dwell on the function $\rho(z)$.

If the moment problem is determinate the series

$$\sum_0^{\infty} | P_k(z) |^2 \qquad\qquad \text{...[2.23]}$$

diverges at any non-real point z and, therefore the function $\rho(z)$ can be different from zero only at points on the real axis.

If on the other hand, the moment problem is indeterminate, the series $[2.23]$ and also the series

$$\sum_0^\infty |Q_k(z)|^2$$

are convergent everywhere. Moreover these series converge uniformly in any finite part of the plane. Therefore in the case of the indeterminate moment problem $\rho(z)$ is everywhere different from zero and continuous. Since $s_0 = 1$ and therefore $P_0(z) = 1$, we have $\rho(z) < 1$ and therefore, everywhere in the complex plane:

$$0 < \rho(z) < 1.$$

THEOREM 2.4.1. (M. RIESZ[3]). *If a moment problem is indeterminate then*

$$\int_{-\infty}^\infty \frac{\ln \dfrac{1}{\rho(u)}}{1+u^2} \, du < \infty. \qquad \qquad ...[2.24]$$

Proof.† By assumption it is possible to find two solutions of the moment problem $\sigma_1(u)$ and $\sigma_2(u)$ for which the difference

$$\Delta = \int_{-\infty}^\infty \frac{d\sigma_1(u)}{u-i} - \int_{-\infty}^\infty \frac{d\sigma_2(u)}{u-i}$$

is not zero. Since for any polynomial $R(u)$ we have

$$\int_{-\infty}^\infty R(u) \, d\sigma_1(u) = \int_{-\infty}^\infty R(u) \, d\sigma_2(u),$$

it follows that

$$\Delta = \int_{-\infty}^\infty \left\{ \frac{1}{u-i} - R(u) \right\} d\sigma_1(u) - \int_{-\infty}^\infty \left\{ \frac{1}{u-i} - R(u) \right\} d\sigma_2(u),$$

whence we have

$$|\Delta| \leq \int_{-\infty}^\infty \left| \frac{1}{u-i} - R(u) \right| d\sigma_1(u) + \int_{-\infty}^\infty \left| \frac{1}{u-i} - R(u) \right| d\sigma_2(u).$$

This inequality with an appropriate choice of the polynomial $R(u)$ leads to a certain estimate of the magnitude of Δ, from which we shall derive the inequality $[2.24]$.

† The proof given here differs from that initially given by M. Riesz, see AKHIEZER[7].

Indeed we shall show immediately that for any n there exists a polynomial $R(u)$ for which we have everywhere on the axis $-\infty < u < \infty$

$$\left|\frac{1}{u-i} - R(u)\right| = \frac{1}{\sqrt{1+u^2}\sqrt{\rho_n(u)}} \exp\left\{-\frac{1}{2\pi}\int_{-\infty}^{\infty} \frac{\ln\frac{1}{\rho_n(t)}}{1+t^2} dt\right\} \quad \ldots[2.25]$$

With the aid of this equation we find that

$$|\Delta| \leq 2\exp\left\{-\frac{1}{2\pi}\int_{-\infty}^{\infty} \frac{\ln\frac{1}{\rho_n(t)}}{1+t^2} dt\right\} \cdot \sup \int_{-\infty}^{\infty} \frac{d\sigma(u)}{\sqrt{1+u^2}\sqrt{\rho_n(u)}},$$

whence

$$\exp\left\{\frac{1}{2\pi}\int_{-\infty}^{\infty} \frac{\ln\frac{1}{\rho_n(t)}}{1+t^2} dt\right\} \leq \frac{2}{|\Delta|} \sup \int_{-\infty}^{\infty} \frac{d\sigma(u)}{\sqrt{1+u^2}\sqrt{\rho_n(u)}}, \quad \ldots[2.26]$$

where the supremum is taken over all solutions of the moment problem while n is fixed. This supremum is easy to estimate. For, given any solution $\sigma(u)$ of the moment problem we have

$$\int_{-\infty}^{\infty} \frac{d\sigma(u)}{\sqrt{1+u^2}\sqrt{\rho_n(u)}} \leq \left\{\int_{-\infty}^{\infty} d\sigma(u)\right\}^{\frac{1}{2}}\left\{\int_{-\infty}^{\infty} \frac{d\sigma(u)}{(1+u^2)\rho_n(u)}\right\}^{\frac{1}{2}}$$

and

$$\int_{-\infty}^{\infty} \frac{d\sigma(u)}{(1+u^2)\rho_n(u)} = \mathrm{Im}\int_{-\infty}^{\infty} \frac{1}{u-i}\sum_0^n [P_k(u)]^2 d\sigma(u)$$

$$= \mathrm{Im}\int_{-\infty}^{\infty} \frac{1}{u-i}\sum_0^n P_k(u)[P_k(u)-P_k(i)] d\sigma(u)$$

$$+ \mathrm{Im}\int_{-\infty}^{\infty} \frac{1}{u-i}\sum_0^n P_k(i)[P_k(u)-P_k(i)] d\sigma(u)$$

$$+ \mathrm{Im}\int_{-\infty}^{\infty} \frac{1}{u-i}\sum_0^n [P_k(i)]^2 d\sigma(u) = I_1+I_2+I_3.$$

The quantity I_1 is zero because each one of the following integrals becomes zero:

$$\int_{-\infty}^{\infty} P_k(u)\frac{P_k(u)-P_k(i)}{u-i} d\sigma(u).$$

Further we have

$$| I_2 | \leq \sum_0^n | P_k(i) | \, | Q_k(i) | \leq \frac{1}{2} \sum_0^n \{ | P_k(i) |^2 + | Q_k(i) |^2 \}$$

and

$$| I_3 | \leq \sum_0^n | P_k(i) |^2 \int_{-\infty}^{\infty} \frac{d\sigma(u)}{\sqrt{1+u^2}} \leq \sum_0^n | P_k(i) |^2 \int_{-\infty}^{\infty} d\sigma(u).$$

By virtue of the indeterminate nature of our moment problem the series

$$\sum_0^{\infty} | P_k(i) | , \quad \sum_0^{\infty} | Q_k(i) |^2$$

converge.   It follows from all that has been said that

$$2 \sup \int_{-\infty}^{\infty} \frac{d\sigma(u)}{\sqrt{1+u^2}\sqrt{\rho_n(u)}} < C,$$

where C is $< \infty$ and independent of n.

With the aid of this estimate the inequality [2.26] may be rewritten in the form

$$\frac{1}{2\pi} \int_{-\infty}^{\infty} \frac{\ln \dfrac{1}{\rho_n(u)}}{1+u^2} \, du < \ln \frac{C}{| \Delta |}.$$

Increasing n to ∞ we find that

$$\frac{1}{2\pi} \int_{-\infty}^{\infty} \frac{\ln \dfrac{1}{\rho(u)}}{1+u^2} \, du = \lim_{n \to \infty} \frac{1}{2\pi} \int_{-\infty}^{\infty} \frac{\ln \dfrac{1}{\rho_n(u)}}{1+u^2} \, du \leq \ln \frac{C}{| \Delta |} < \infty,$$

and the proof of the theorem is complete.

It remains to show how to construct the polynomial $R(u)$ which satisfies the equation [2.25].

We take the polynomial

$$\frac{1}{\rho_n(u)} = \sum_0^n [P_k(u)]^2 \quad (-\infty < u < \infty)$$

of degree $2n$, which is positive and therefore can be represented in the form

$$\frac{1}{\rho_n(u)} = | S_n(u) |^2 \quad (-\infty < u < \infty),$$

where

$$S_n(u) = \frac{1}{\sqrt{\rho_n(0)}} \prod_{k=1}^{n} \left(1 - \frac{u}{c_k}\right) \quad (\mathrm{Im}\ c_k < 0).$$

Having found $S_n(u)$, we put

$$\frac{S_n(u)}{S_n(i)} = 1 - (u-i)R(u). \qquad \qquad \ldots[2.27]$$

Thus $R(u)$ is a polynomial of degree $n-1$. It follows from $[2.27]$ that

$$\frac{1}{u-i} - R(u) = \frac{1}{u-i} \frac{S_n(u)}{S_n(i)}$$

so that

$$\left| \frac{1}{u-i} - R(u) \right| = \frac{1}{\sqrt{1+u^2}\sqrt{\rho_n(u)}} \frac{1}{|S_n(i)|} \quad (-\infty < u < \infty).$$

To obtain equation $[2.25]$ it remains to express the last factor in terms of $\rho_n(t)$. This can be done with the aid of the equation

$$\ln |S_n(i)| = \frac{1}{\pi} \int_{-\infty}^{\infty} \frac{\ln |S_n(t)|}{1+t^2}\,dt = \frac{1}{2\pi} \int_{-\infty}^{\infty} \frac{\ln \dfrac{1}{\rho_n(t)}}{1+t^2}\,dt,$$

which is a particular case of the Poisson integral but can be verified directly by elementary methods.†

4.2. Let us remind ourselves of the four families of polynomials which we introduced in Section **2.4.** of Chapter 1. They were defined by means of the formulae

$$A_n(z) = z \sum_{0}^{n-1} Q_k(0)Q_k(z),$$

$$B_n(z) = -1 + z \sum_{0}^{n-1} Q_k(0)P_k(z),$$

$$C_n(z) = 1 + z \sum_{0}^{n-1} P_k(0)Q_k(z),$$

$$D_n(z) = z \sum_{0}^{n-1} P_k(0)P_k(z).$$

---

† It is sufficient to prove (for instance by differentiating with respect to the parameter and the theory of residues) that

$$\frac{1}{\pi} \int_{-\infty}^{\infty} \frac{\ln \left| 1 - \dfrac{t}{c} \right|}{1+t^2}\,dt = \ln \left| 1 - \frac{i}{c} \right| \quad (\mathrm{Im}\ c < 0).$$

We shall assume further that the moment problem we are discussing is indeterminate. Then the series

$$\sum_0^\infty |P_k(z)|^2, \quad \sum_0^\infty |Q_k(z)|^2$$

converge, uniformly in fact, in any finite part of the complex plane. It follows therefore from the Cauchy-Bunyakovskii inequality that for $n \to \infty$ the polynomials $A_n(z)$, $B_n(z)$, $C_n(z)$ and $D_n(z)$ tend uniformly to their limits in every finite part of the plane. Therefore these limits are entire transcendental functions

$$
\left.
\begin{aligned}
A(z) &= z \sum_0^\infty Q_k(0)Q_k(z), \\[1em]
B(z) &= -1 + z \sum_0^\infty Q_k(0)P_k(z), \\[1em]
C(z) &= 1 + z \sum_0^\infty P_k(0)Q_k(z), \\[1em]
D(z) &= z \sum_0^\infty P_k(0)P_k(z).
\end{aligned}
\right\}
\qquad \ldots[2.28]
$$

They are related by the equation

$$A(z)D(z) - B(z)C(z) = 1, \qquad \ldots[2.29]$$

which is a consequence of the relation [1.24].

We also recall equation [1.25] which has the form

$$w_n(z) = -\frac{A_n(z)t - C_n(z)}{B_n(z)t - D_n(z)}, \qquad \ldots[2.30a]$$

where t is a real parameter. For every value of this parameter we have

$$w_n(z) = \int_{-\infty}^\infty \frac{d\sigma_n(u)}{u - z} \quad (\text{Im } z \neq 0), \qquad \ldots[2.30b]$$

where $\sigma_n(u)$ is the solution of a certain truncated moment problem. Going to the limit $n \to \infty$ for fixed t in equation [2.30a] and applying Helly's theorem to equation [2.30b] we find that

$$-\frac{A(z)t - C(z)}{B(z)t - D(z)} = w(z) \qquad \ldots[2.31]$$

and

$$w(z) = \int_{-\infty}^\infty \frac{d\sigma(u)}{u - z} \quad (\text{Im } z \neq 0), \qquad \ldots[2.23]$$

where $\sigma(u)$ is a certain solution of the full moment problem and $w(z)$ a certain meromorphic function. To bring out that $\sigma(u)$ and $w(z)$ depend on the choice of the value of the parameter t we shall sometimes use the notation $\sigma_t(u)$ and $w(z\,;\,t)$.

For fixed z (Im $z \neq 0$) the point $w(z\,;\,t)$ describes the limiting circle $K_\infty(z)$ when t varies along the real axis. Therefore the solution $\sigma(u)$ of the moment problem which appears in equation $[2.32]$ is N-extremal. It follows from the representation $[2.32]$ of the function $[2.31]$ that for any real t

$$\mathrm{Im}\left\{-\frac{A(z)t - C(z)}{B(z)t - D(z)}\right\}: \mathrm{Im}\ z > 0 \quad (\mathrm{Im}\ z \neq 0),$$

and also that for any real t the zeros of the functions

$$A(z)t - C(z), \quad B(z)t - D(z)$$

are real, simple and interlace.†

4.3. DEFINITION **2.4.3.** We shall use the name *Nevanlinna matrix* for any matrix

$$\left\|\begin{array}{cc} a(z) & c(z) \\ b(z) & d(z) \end{array}\right\|,$$

the elements of which are entire transcendental functions, if the following conditions are satisfied:

a) The identity holds:

$$a(z)\,d(z) - b(z)\,c(z) = 1;$$

b) For any fixed real $t(-\infty < t < \infty)$ the function

$$w(z) = \frac{a(z)t - c(z)}{b(z)t - d(z)}$$

is regular in the half-planes Im $z > 0$ and Im $z < 0$ and satisfies the inequality

$$\frac{\mathrm{Im}\ w(z)}{\mathrm{Im}\ z} > 0.$$

† The fact that the zeros of the denominator $B(z)t - D(z)$ are real follows from the regularity of the function $w(z)$ for Im $z \neq 0$. With the aid of the Stieltjes-Perron inversion formula we find that in every interval of the real axis which does not contain any zero of the denominator the function $\sigma(u)$ is constant. Therefore the representation $[2.32]$ is in fact of the form

$$w(z) = \sum_k \frac{\mu_k}{\lambda_k - z} \quad (\mu_k > 0),$$

whence the properties of the zeros of numerator and denominator of $[2.31]$ can be obtained without difficulty.

It is easy to see that the elements of the Nevanlinna matrix must be real entire functions.

The matrix

$$\left\| \begin{array}{cc} A(z) & C(z) \\ B(z) & D(z) \end{array} \right\|$$

belonging to an indeterminate moment problem is a Nevanlinna matrix. The following theorem shows that it possesses special properties which an arbitrary Nevanlinna matrix need not have.†

THEOREM 2.4.3. (M. RIESZ[3]) *The elements of the Nevanlinna matrix corresponding to an indeterminate moment problem are entire functions of at most the minimal type of order* 1.

The function $\rho(z)$ *corresponding to an indeterminate moment problem satisfies the inequality*

$$\frac{1}{\rho(z)} \leqq N_\varepsilon e^{\varepsilon |z|},$$

where ε *is an arbitrary positive number and* N_ε *is independent of* z.

Proof. For any non-real z the points

$$-\frac{A_n(z)}{B_n(z)}, \quad -\frac{C_n(z)}{D_n(z)}$$

lie on the circular circumference $K_n(z)$. Therefore the distance between these points is no greater than the diameter of this circle, i.e.

$$\left| \frac{A_n(z)}{B_n(z)} - \frac{C_n(z)}{D_n(z)} \right| \leqq \frac{\rho_{n-1}(z)}{|y|} \quad (y = \text{Im } z).$$

Hence by virtue of the equation

$$A_n(z)D_n(z) - B_n(z)C_n(z) = 1,$$

it follows that

$$\frac{1}{|B_n(z)D_n(z)|} \leqq \frac{\rho_{n-1}(z)}{|y|}. \qquad \qquad ...[2.33]$$

We note further that from the relations [2.30a] and [2.30b] (first for $t = \infty$ and then for $t = 0$) the following inequalities result

$$\frac{|A_n(z)|}{|B_n(z)|} < \frac{1}{|y|}, \quad \frac{|C_n(z)|}{|D_n(z)|} < \frac{1}{|y|} \quad (s_0 = 1). \qquad ...[2.33a]$$

† For instance, these properties are not possessed by the Nevanlinna matrix

$$\left\| \begin{array}{cc} \cos z & -\sin z \\ \sin z & \cos z \end{array} \right\|.$$

On the other hand we have

$$|B_n(z)| \leq 1 + |z| \sqrt{\sum_0^\infty [Q_k(0)]^2} \sqrt{\sum_0^\infty |P_k(z)|^2} \leq 1 + M\frac{|z|}{\sqrt{\rho(z)}}$$

and similarly

$$|D_n(z)| \leq \frac{1}{\sqrt{\rho(0)}}\frac{|z|}{\sqrt{\rho(z)}} < 1 + M\frac{|z|}{\sqrt{\rho(z)}},$$

where

$$M = \sqrt{\sum_0^\infty \{[Q_k(0)]^2 + [P_k(0)]\}^2}.$$

Since $B_n(z)$ is a real quasiorthogonal polynomial, its zeros are real and therefore, according to Poissons formula† we have

$$\ln|B_n(x+iy)| = \frac{|y|}{\pi}\int_{-\infty}^\infty \frac{\ln|B_n(\xi)|}{(x-\xi)^2+y^2}\,d\xi \quad (y \neq 0).$$

Therefore

$$\ln|B_n(x+iy)| \leq \frac{|y|}{\pi}\int_{-\infty}^\infty \frac{\ln\left\{1+M\frac{|\xi|}{\sqrt{\rho(\xi)}}\right\}}{(x-\xi)^2+y^2}\,d\xi,$$

where the existence of the integral on the right hand side is assured by Theorem **2.4.1.** The function $\ln|D_n(x+iy)|$ satisfies an exactly similar inequality.

We put

$$\phi(\xi) = \ln\left\{1+M\frac{|\xi|}{\sqrt{\rho(\xi)}}\right\},$$

so that $\phi(\xi) \geq 0$ and

$$\int_{-\infty}^\infty \frac{\phi(\xi)}{1+\xi^2}\,d\xi < \infty.$$

We must examine the behaviour of the integral

$$\frac{1}{\pi}\int_{-\infty}^\infty \frac{\phi(\xi)}{(x-\xi)^2+y^2}\,d\xi$$

for $z = x+iy \to \infty$. This can be done very simply for the arguments

$$\delta \leq \arg z \leq \pi-\delta, \quad \pi+\delta \leq \arg z \leq 2\pi-\delta, \qquad \ldots[2.34]$$

† See footnote on p. 53.

where the quantity δ $(0 < \delta < \frac{1}{2}\pi)$ is fixed. Indeed, for such arguments and for $|\xi| \geqq 1$ the inequality holds:

$$(x-\xi)^2 + y^2 = |z - \xi|^2 \geqq \xi^2 \sin^2 \delta \geqq \frac{1+\xi^2}{2} \sin^2 \delta.$$

Therefore, for $T > 1$ we have

$$\frac{1}{\pi}\int_{-\infty}^{-T} + \frac{1}{\pi}\int_{T}^{\infty} \frac{\phi(\xi)}{(x-\xi)^2 + y^2}\,d\xi \leqq \frac{2}{\pi \sin^2 \delta}\left[\int_{-\infty}^{-T} + \int_{T}^{\infty}\frac{\phi(\xi)}{1+\xi^2}\,d\xi\right],$$

and the right hand side may be made as small as required by taking a sufficiently large value of T. Then, with fixed T, we get

$$\lim_{|y|\to\infty} \frac{1}{\pi}\int_{-T}^{T}\frac{\phi(\xi)}{(x-\xi)^2+y^2}\,d\xi = 0.$$

It follows from what has been said that for the range of arguments [2.34] the inequalities

$$|B_n(z)| \leqq M_\varepsilon e^{\varepsilon|z|}, \quad |D_n(z)| \leqq M_\varepsilon e^{\varepsilon|z|} \qquad \ldots[2.35]$$

hold, where M_ε is independent of z and of n, and ε is an arbitrarily chosen positive number.

Since the roots of the polynomials $B_n(z)$ and $D_n(z)$ are real, the functions $|B_n(x+iy)|$, $|D_n(x+iy)|$ increase monotonically with $|y|$. Therefore it follows that in the complementary range of arguments

$$-\delta \leqq \arg z \leqq \delta, \quad \pi - \delta \leqq \arg z \leqq \pi + \delta$$

the inequalities hold:

$$|B_n(x+iy)| \leqq |B_n(x[1 + i\tan\delta])|,$$

$$|D_n(x+iy)| \leqq |D_n(x[1 + i\tan\delta])|.$$

By virtue of [2.35] it follows therefore that for the complementary arguments one has

$$|B_n(z)| \leqq M_\varepsilon e^{\frac{\varepsilon}{\cos\delta}|x|} \leqq M_\varepsilon e^{\frac{\varepsilon}{\cos\delta}|z|},$$

$$|D_n(z)| \leqq M_\varepsilon e^{\frac{\varepsilon}{\cos\delta}|x|} \leqq M_\varepsilon e^{\frac{\varepsilon}{\cos\delta}|z|}.$$

The same inequalities are also valid in the argument range [2.34]. Since $\varepsilon > 0$ is arbitrary the inequalities may be rewritten in the form [2.35] (possibly with a different coefficient M_ε). Thus one can take it that the inequalities [2.35] are valid over the whole z-plane.

As a result of [2.33a] the following inequalities hold in the whole z-plane.

$$| A_n(x+iy) | < \frac{1}{|y|} M_\varepsilon e^{\varepsilon |z|}, \quad | C_n(x+iy) | < \frac{1}{|y|} M_\varepsilon e^{\varepsilon |z|}.$$

For $| y | \geqq 1$ these inequalities may be replaced by the following

$$| A_n(x+iy) | \leqq M_\varepsilon e^{\varepsilon |z|}, \quad | C_n(x+iy) | \leqq M_\varepsilon e^{\varepsilon |z|}. \quad ...[2.36]$$

And since the functions $| A_n(x+iy) |$ and $| C_n(x+iy) |$ increase monotonically with increasing $| y |$ we have that

$$| A_n(x+iy) | \leqq | A_n(x+i) | \leqq M_\varepsilon e^{\varepsilon |x+i|} \leqq M e^\varepsilon e^{\varepsilon |z|},$$

if $0 \leqq y \leqq 1$, and

$$| A_n(x+iy) | \leqq | A_n(x-i) | \leqq M_\varepsilon e^{\varepsilon |x-1|} \leqq M_\varepsilon e^\varepsilon e^{\varepsilon |z|},$$

if $-1 \leqq y \leqq 0$. Similarly, we have for $| y | \leqq 1$ that

$$| C_n(x+iy) | \leqq M_\varepsilon e^\varepsilon e^{\varepsilon |z|}.$$

Therefore the inequalities [2.36] (possibly with a different coefficient M_ε) hold over the whole z plane. Passing to the limit $n \to \infty$ in [2.35] and [2.36] we find that

$$| A(z) | \leqq M_\varepsilon e^{\varepsilon |z|} \quad | B(z) | \leqq M_\varepsilon e^{\varepsilon |z|},$$
$$| C(z) | \leqq M_\varepsilon e^{\varepsilon |z|} \quad | D(z) | \leqq M_\varepsilon e^{\varepsilon |z|},$$

and therefore the first statement of our theorem is proved.

To prove the second statement we take the inequalities [2.33] and [2.35]. From these it follows that

$$\frac{1}{\rho_{n-1}(z)} \leqq \frac{1}{|y|} | B_n(z)D_n(z) | \leqq \frac{1}{|y|} N_\varepsilon e^{\varepsilon |z|}.$$

For $| y | \geqq 1$ the factor $1/| y |$ may be omitted and if $| y | \leqq 1$, one can, as above, use the fact that the function

$$\frac{1}{\rho_{n-1}(z)} = \sum_0^{n-1} | P_k(x+iy) |^2$$

increases monotonically with increasing $| y |$. As a result we find that over the whole z plane:

$$\frac{1}{\rho_{n-1}(z)} \leqq N_\varepsilon e^{\varepsilon |z|},$$

where N_ε is independent of n and of z. It now remains to perform the passage to the limit $n \to 0$.

5. Extremal properties of the functions $\rho_n(z)$ and $\rho(z)$

5.1. From the definition of the function $\rho_n(z)$ given in Section **4.1.** of this chapter it is evident that its construction requires the knowledge of the orthogonal polynomials $P_k(z)$ only up to the n-th included, or in other words the knowledge of the moments s_k only up to s_{2n} inclusive. Therefore in questions relating to the $\rho_n(z)$ we shall assume that a finite sequence $\{s_k\}_0^{2n}$ is given and shall assume that this sequence is positive, which amounts to saying that the form

$$\sum_{i,\,k=0}^{n} s_{i+k} x_i x_k$$

is positive. If such a sequence is given, it means that a positive functional S is given in the space of all polynomials of degree $\leq 2n$. We note also that if, in addition to the positive sequence $\{s_k\}_0^{2n}$ an arbitrary real number s_{2n+1} is also given one can construct the polynomial

$$\Delta_{n+1}(\lambda) = \begin{vmatrix} 1 & \lambda & \lambda^2 & \dots & \lambda^{n+1} \\ s_0 & s_1 & s_2 & \dots & s_{n+1} \\ \dots & \dots & \dots & \dots & \dots \\ \dots & \dots & \dots & \dots & \dots \\ s_n & s_{n+1} & s_{n+2} & \dots & s_{2n+1} \end{vmatrix},$$

which satisfies the orthogonality relations

$$S\{\Delta_{n+1}(\lambda)\lambda^k\} = 0 \quad (k = 0, 1, \dots, n)$$

and can often serve to replace the " non-existent " polynomial $P_{n+1}(\lambda)$

THEOREM **2.5.1.** *If S is a positive functional in the space of all polynomials of degree $\leq n$ and if $\mathfrak{U}_n$ denotes the set of all polynomials $R_n(\lambda)$ of degree $\leq n$ which satisfy the condition*

$$R_n(z) = 1, \qquad \dots[2.37]$$

where z is a given number (real or complex) then

$$\min_{R_n \in \mathfrak{U}_n} S\{|R_n(\lambda)|^2\} = \rho_n(z),$$

and the polynomial $R_n^{\text{extr}}(\lambda)$ on which this minimum is reached is equal to

$$R_n^{\text{extr}}(\lambda) = \frac{\sum_0^n \overline{P_k(z)} P_k(\lambda)}{\sum_0^n |P_k(z)|^2} = \frac{h_n(\bar{z}, \lambda)}{h_n(\bar{z}, z)}.$$

Proof. Any polynomial $R_n(\lambda)$ of degree $\leqq n$ which satisfies condition [2.37] can be represented in the form

$$R_n(\lambda) = A_0 P_0(\lambda) + A_1 P_1(\lambda) + \ldots + A_n P_n(\lambda), \qquad \ldots [2.38]$$

where the coefficients A_k are related by the equation

$$A_0 P_0(z) + A_1 P_1(z) + \ldots + A_n P_n(z) = 1. \qquad \ldots [2.39]$$

Since it follows from [2.38] that

$$\mathfrak{S}\{|R_n(\lambda)|^2\} = |A_0|^2 + |A_1|^2 + \ldots + |A_n|^2,$$

we are discussing the minimum of the quantity

$$|A_0|^2 + |A_1|^2 + \ldots + |A_n|^2$$

under the condition [2.39]. But it follows from this condition that

$$1 \leqq \{|A_0|^2 + |A_1|^2 + \ldots + |A_n|^2\}\{|P_0(z)|^2 + |P_1(z)|^2 + \ldots + |P_n(z)|^2\},$$

where the equality holds precisely if

$$A_k = \frac{\overline{P_k(z)}}{\sum\limits_0^n |P_j(z)|^2} \qquad (k = 0, 1, 2, \ldots, n)$$

and only then. From what has been said it follows both that

$$\min \mathfrak{S}\{|R_n(\lambda)|^2\} = \frac{1}{\sum\limits_0^n |P_j(z)|^2} = \rho_n(z)$$

and that

$$R_n^{\text{extr}}(\lambda) = \frac{\sum\limits_0^n \overline{P_k(z)} P_k(\lambda)}{\sum\limits_0^n |P_k(z)|^2}.$$

5.2. We shall denote by $\mathfrak{M}_m$ the set of all (non-decreasing) solutions $\sigma(u)$ of the truncated moment problem

$$\int_{-\infty}^{\infty} u^k \, d\sigma(u) = s_k \quad (k = 0, 1, \ldots, m),$$

where $\{s_k\}_0^m$ is a given positive sequence. In essence this means that the sequence $\{s_k\}_0^{2\left[\frac{m}{2}\right]}$ is positive and s_m, if m is odd, is an arbitrary real number.

THEOREM **2.5.2.** a) *If the sequence* $\{s_k\}_0^{2n}$ *is positive and if* x *is an arbitrary real number then*

$$\max_{\sigma(u) \,\in\, \mathfrak{M}_{2n}} \{\sigma(x+0) - \sigma(x-0)\} \leqq \rho_n(x) \qquad \ldots[2.40]$$

and for $P_n(x) \neq 0$ *the equality sign holds in equation* [2.40].

b) *If the sequence* $\{s_k\}_0^{2n-1}$ *is positive and the real number* x *is such that* $\Delta_n(x) = 0$ *then*

$$\max_{\sigma(u) \,\in\, \mathfrak{M}_{2n-1}} \{\sigma(x+0) - \sigma(x-0)\} = \rho_{n-1}(x).$$

Thus the quantity $\rho_n(x)$ *for any real value of* x *represents the maximum mass which can be concentrated at the point* x *under the condition that a definite number of moments is given.*

Proof. We start from statement b). Using the considerations of Sections **4.1** and **4.2** in Chapter 1, and noting that in place of the polynomial $P_n(\lambda)$ we can operate with the polynomial $\Delta_n(\lambda)$, we can find the representation

$$s_k = \sum_{i=1}^{n} \mu_i \lambda_i^k \quad (k = 0, 1, \ldots, 2n-1), \qquad \ldots[2.41]$$

where $\mu_i = \rho_{n-1}(\lambda_i)$ and λ_i is a zero of the polynomial $\Delta_n(\lambda)$. One of the zeros λ_i is equal to x. Thus it follows from the representation [2.41] that

$$\max_{\sigma(u) \,\in\, \mathfrak{M}_{2n-1}} \{\sigma(x+0) - \sigma(x-0)\} \geqq \rho_{n-1}(x),$$

i.e. the maximum mass under discussion is not less than $\rho_{n-1}(x)$. It remains to prove that this maximum mass is not greater than $\rho_{n-1}(x)$. To do this we take any function $\sigma(u) \in \mathfrak{M}_{2n-1}$ and write the obvious inequality

$$\int_{-\infty}^{\infty} |R(u)|^2 \, d\sigma(u) \geqq |R(x)|^2 \{\sigma(x+0) - \sigma(x-0)\}, \quad \ldots[2.42]$$

where $R(u)$ is an arbitrary polynomial of degree $n-1$. We confine ourselves to polynomials $R(u)$ for which $R(x) = 1$ and take that one which minimizes the left hand side. Using Theorem **2.5.1** we find that

$$\rho_{n-1}(x) \geqq \sigma(x+0) - \sigma(x-0).$$

This proves statement b).

We now examine statement a). If $P_n(x) = 0$, then $\rho_{n-1}(x) = \rho_n(x)$ and because of statement b) we have

$$\max_{\sigma(u)\,\in\,\mathfrak{M}_{2n}} \{\sigma(x+0)-\sigma(x-0)\} \leqq \max_{\sigma(u)\,\in\,\mathfrak{M}_{2n-1}} \{\sigma(x+0)-\sigma(x-0)\}$$

$$= \rho_{n-1}(x) = \rho_n(x).$$

If on the other hand $P_n(x) \neq 0$ it is possible to determine s_{2n+1} in such a way, and uniquely, that $\Delta_{n+1}(x) = 0$, and then we have, from statement b)

$$\max_{\sigma(u)\,\in\,\mathfrak{M}_{2n}} \{\sigma(x+0)-\sigma(x-0)\} \geqq \max_{\sigma(u)\,\in\,\mathfrak{M}_{2n+1}} \{\sigma(x+0)-\sigma(x-0)\} = \rho_n(x).$$

On the other hand, using inequality [2.42] and Theorem 2.5.1 we can establish, as above, that

$$\max_{\sigma(u)\,\in\,\mathfrak{M}_{2n}} \{\sigma(x+0)-\sigma(x-0)\} \leqq \rho_n(x).$$

Thus statement a) is also proved.

5.3. THEOREM 2.5.3. *The maximum mass that can be concentrated at a given point x ($-\infty < x < \infty$) in a distribution of mass on the axis which corresponds to a solution of the full moment problem, is $\rho(x)$.*

Proof. We denote the required maximum mass by μ. Since the solution of the full moment problem is also a solution of all truncated problems we have by virtue of Theorem 2.5.2 that

$$\mu \leqq \rho_n(x)$$

for any n. Therefore

$$\mu \leqq \rho(x). \qquad\qquad [2.43]$$

On the other hand, by virtue of Theorem 2.5.2 we can find an infinite sequence $\{\sigma_{n_i}(u)\}_{i=1}^{\infty}$ of solutions for the truncated problems of arbitrarily high order such that

$$\sigma_{n_i}'(x+0)-\sigma_{n_i}'(x-0) = \rho_{n_i}(x) \quad [\geqq \rho(x)].$$

Using Helly's theorem we construct the sequence $\{\sigma_{n_i}'(u)\}_{i=1}^{\infty}$ and the function $\sigma(u)$ in such a way that at all points of continuity of $\sigma(u)$ we have

$$\lim_{i\to\infty} \sigma_{n_i}'(u) = \sigma(u).$$

Assume that $\alpha < x$ and $\beta > x$ are points of continuity of $\sigma(u)$. In that case

$$\sigma(\beta)-\sigma(\alpha) = \lim_{i\to\infty} \{\sigma_{n_i}'(\beta)-\sigma_{n_i}'(\alpha)\} \geqq \rho(x),$$

since

$$\sigma_{n_i}'(\beta) - \sigma_{n_i}'(\alpha) \geqq \sigma_{n_i}'(x+0) - \sigma_{n_i}'(x-0) \geqq \rho(x),$$

whence it follows that

$$\sigma(x+0) - \sigma(x-0) \geqq \rho(x).$$

Therefore

$$\mu \geqq \rho(x).$$

Comparing this inequality with [2.43] we find precisely that the maximum mass is $\rho(x)$.

COROLLARY 2.5.3. *If* $\rho(x) \neq 0 (-\infty < x < \infty)$ *the moment problem is indeterminate.*

Indeed, if the moment problem were determinate the function $\rho(x)$ would represent the values of concentrated mass at points x, *all for one solution*, while the set of points of discontinuity of a monotonic function is denumerable.

5.4. In the Foreword to this book we gave the statement of the problem due to Chebyshev concerning the limiting values of some integrals. A large number of papers has been devoted to this problem and its generalizations,† and it is impossible within the scope of this book to give any detailed exposition of results relating to this.

We shall confine ourselves to stating one proposition which is basic and typical. This proposition was essentially formulated (in different terms, of course) by CHEBYSHEV[1] himself; it was then proved by MARKOV[3] and independently by STIELTJES[1].

THEOREM 2.5.4. *Assume that the sequence* $\{s_k\}_0^{2n-1}$ *is positive and that* $\sigma(u) \in \mathfrak{M}_{2n-1}$.

In that case we have for $i = 1, 2, \ldots, n-1$ *that*

$$\int_{-\infty}^{\lambda_i+0} d\sigma(u) \leqq \rho_{n-1}(\lambda_1) + \rho_{n-1}(\lambda_2) + \ldots + \rho_{n-1}(\lambda_i) \leqq \int_{-\infty}^{\lambda_{i+1}-0} d\sigma(u),$$

$$\ldots [2.44]$$

where $\lambda_1, \lambda_2, \ldots, \lambda_n$ *are the zeros of the polynomial* $\Delta_n(\lambda)$.

If in addition it is known that $\sigma(u)$ *has more than n points of increase then both the* $\leqq$ *signs in relation* [2.43] *can be replaced by* < *signs and the inequality following holds:*

$$0 < \int_{-\infty}^{\lambda_1-0} d\sigma(u), \quad \int_{-\infty}^{\lambda_n+0} d\sigma(u) < s_0. \qquad \ldots [2.44a]$$

† The reader can find a modern and detailed exposition of the question in the revue by KREIN[9].

Proof. We construct a polynomial $R(u)$ of degree $2n-2$ from the following conditions which determine it uniquely:

$$R(\lambda_1) \;\; = 1, \quad R'(\lambda_1) \;\; = 0,$$
$$R(\lambda_2) \;\; = 1, \quad R'(\lambda_2) \;\; = 0,$$
$$\cdots\cdots\cdots\cdots\cdots\cdots$$
$$R(\lambda_{i-1}) = 1, \quad R'(\lambda_{i-1}) = 0,$$
$$R(\lambda_i) \;\; = 1,$$
$$R(\lambda_{i+1}) = 0, \quad R'(\lambda_{i+1}) = 0,$$
$$\cdots\cdots\cdots\cdots\cdots\cdots$$
$$R(\lambda_n) \;\; = 0, \quad R'(\lambda_n) \;\; = 0.$$

Fig. 1 represents schematically the graph of the polynomial $R(u)$. It is essential that for $u \leqq \lambda_i$ the polynomial $R(u)$ satisfies the inequality

$$R(u) \geqq 1,$$

and that for $u > \lambda_i$ we have in any case

$$R(u) \geqq 0.$$

These properties of the polynomial $R(u)$ follow from the fact that by Rolle's theorem $R'(u)$ has at least one zero in each open interval

$$(\lambda_1, \lambda_2), \;\; (\lambda_2, \lambda_3), \;\; ..., \;\; (\lambda_{i-1}, \lambda_i), \;\; (\lambda_{i+1}, \lambda_{i+2}), \;\; ..., \;\; (\lambda_{n-1}, \lambda_n).$$

Indeed we have here $n-2$ zeros in all. In addition $R'(u)$ becomes zero at all points λ_k apart from λ_i, which gives us $n-1$ zeros. Thus we have obtained $2n-3$ zeros and therefore $R'(u)$ has no other zeros. Hence it follows that in the interval $(\lambda_i, \lambda_{i+1})$ the polynomial $R(u)$ decreases

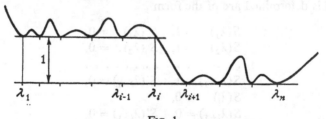

FIG. 1

monotonically and also that the points λ_k, with the exception of λ_i, are the only positions of relative minima of $R(u)$. The further discussion is based on the application of the quadrature formula which here has the form

$$\int_{-\infty}^{\infty} R(u)\, d\sigma(u) = \sum_1^n \rho_{n-1}(\lambda_k) R(\lambda_k).$$

The right hand side of this expression is equal to

$$\rho_{n-1}(\lambda_1)+\rho_{n-1}(\lambda_2)+ \ \cdots \ +\rho_{n-1}(\lambda_i).$$

On the other hand the left hand side satisfies the inequality

$$\int_{-\infty}^{\infty} R(u)\, d\sigma(u) \geqq \int_{-\infty}^{\lambda_i+0} d\sigma(u),$$

and the $\geqq$ sign can be replaced by the $>$ sign if $\sigma(u)$ has more than n points of increase.

Comparing these relations, we obtain the first part of the statement [2.44] which is to be proved.

The proof of the second part is similar. One must only construct the polynomial differently; we now denote it by $S(u)$. Its graph is

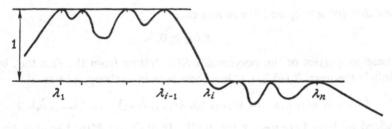

Fig. 2

given schematically in Fig. 2 and the conditions by which this polynomial is determined are of the form

$$
\begin{aligned}
S(\lambda_1) \ &= 1, \quad S'(\lambda_1) \ = 0,\\
S(\lambda_2) \ &= 1, \quad S'(\lambda_2) \ = 0,\\
&\cdot \cdot \cdot \cdot \cdot \cdot \cdot \cdot \cdot \cdot \cdot \cdot \cdot \cdot\\
S(\lambda_{i-1}) &= 1, \quad S'(\lambda_{i-1}) = 0,\\
S(\lambda_i) \ &= 0,\\
S(\lambda_{i+1}) &= 0, \quad S'(\lambda_{i+1}) = 0,\\
&\cdot \cdot \cdot \cdot \cdot \cdot \cdot \cdot \cdot \cdot \cdot \cdot \cdot \cdot\\
S(\lambda_n) \ &= 0, \quad S'(\lambda_n) \ = 0.
\end{aligned}
$$

We need not dwell on the proof of the inequalities [2.44a].

COROLLARY 2.5.4. *If two non-decreasing functions $\sigma_1(u)$ and $\sigma_2(u)$ satisfy the relations*

$$s_k = \int_{-\infty}^{\infty} u^k\, d\sigma_1(u) = \int_{-\infty}^{\infty} u^k\, d\sigma_2(u) \quad (k = 0, 1, 2, ..., 2n-2), \quad ...[2.45]$$

where $\{s_k\}_0^{2n-1}$ *is a positive sequence, then we have for arbitrary real* x:

$$\left| \int_{-\infty}^{x+0} d\sigma_1(u) - \int_{-\infty}^{x-0} d\sigma_2(u) \right| \leqq \rho_{n-1}(x). \quad \dots [2.46]$$

Proof. We assume to begin with that x is not equal to one of the zeros of the polynomial $P_{n-1}(\lambda)$. In that case we can determine s_{2n-1} in such a way that $\Delta_n(x) = 0$. Therefore x will be one of the zeros $\lambda_1, \lambda_2, \dots, \lambda_n$ of the polynomial $\Delta_n(\lambda)$, say $x = \lambda_i$. Therefore we can write down the inequality†

$$\int_{-\infty}^{x+0} d\sigma_1(u) \leqq \rho_{n-1}(\lambda_1) + \rho_{n-1}(\lambda_2) + \dots + \rho_{n-1}(\lambda_i),$$

$$\int_{-\infty}^{x-0} d\sigma_2(u) \geqq \rho_{n-1}(\lambda_1) + \rho_{n-1}(\lambda_2) + \dots + \rho_{n-1}(\lambda_{i-1}),$$

whence it follows that

$$\int_{-\infty}^{x+0} d\sigma_1(u) - \int_{-\infty}^{x-0} d\sigma_2(u) \leqq \rho_{n-1}(\lambda_i) = \rho_{n-1}(x).$$

In order to obtain the inequality [2.46] we must also interchange $\sigma_1(u)$ and $\sigma_2(u)$.

Assume now that x is one of the zeros of the polynomial $P_{n-1}(\lambda)$. In that case we can apply directly the result previously obtained, replacing n by $n-1$ in the condition [2.45]. Therefore the inequality holds:

$$\left| \int_{-\infty}^{x+0} d\sigma_1(u) - \int_{-\infty}^{x-0} d\sigma_2(u) \right| \leqq \rho_{n-2}(x),$$

but this is the same as inequality [2.46], because for $P_{n-1}(x) = 0$ the quantities $\rho_{n-2}(x)$ and $\rho_{n-1}(x)$ are equal.

We note that by virtue of the inequalities of Theorem 2.5.4 the function $\sigma(u)$, having more than n points of increase, cannot remain at a constant value in any of the intervals between two neighbouring points λ_j.

Hence in particular the following proposition follows, which was known already to Stieltjes: *Between any two neighbouring zeros of the orthogonal polynomial $P_m(\lambda)$ there is at least one zero of each orthogonal polynomial $P_n(\lambda)$ of higher degree.*

† If $i = 1$ the second inequality must be replaced by
$$\int_{-\infty}^{x-0} d\sigma_2(u) \geqq 0.$$

6. M. Riesz' method

6.1. If a moment problem is solved and the non-decreasing function $\sigma(u)$ is found, then the functional S which is initially defined on the set of all polynomials by means of the identification

$$S\{u^k\} = s_k \quad (k = 0, 1, 2, \ldots)$$

and by use of algebraic linearity, proves capable of extension by means of the formula

$$S\{f(u)\} = \int_{-\infty}^{\infty} f(u) \, d\sigma(u)$$

to the wider space of all sectionally continuous functions which for $\pm u \to \infty$ increase no more rapidly than a polynomial. With this extension the basic property of the functional is maintained, namely its positive (more precisely non-negative) character, because for any function $f(u) \geqq 0$ of the space just described we have

$$\int_{-\infty}^{\infty} f(u) \, d\sigma(u) \geqq 0.$$

Conversely, if by any method an extension of the functional S has been performed, which maintains the positive character of functions in the space in question, then in particular the functional S becomes meaningful for the function which is equal to unity for $u \leqq t$ and equal to zero for $u > t$, whatever the value of the real number t. The value of the functional S for this function will just be the solution $\sigma(t)$ of the moment problem under discussion.

Thus the moment problem is included in functional analysis as a certain problem concerning the extension of a functional. This approach to the moment problem has its beginnings in the work of F. Riesz and is the basis of the extremely beautiful method due to M. Riesz[3], the exposition of which is the subject of the present section. The method of M. Riesz has great generality and is applicable not only to the power moment problem but also to many related problems. The functional basis of this method is a general theorem, which is discussed in the following section.

6.2. Assume that a certain real linear space E is given and also a linear manifold $\mathfrak{M} \subset E$. Assume in addition that a certain convex cone K is given in E, i.e. a convex manifold which together with an arbitrary element x contains the whole ray αx ($\alpha \geqq 0$).

We shall call an additive and homogeneous functional $\mathfrak{p}\{x\}$, defined in $\mathfrak{M}$, $\mathfrak{K}$-positive† if for any element $x \in \mathfrak{K} \cap \mathfrak{M}$ the inequality $\mathfrak{P}\{x\} \geqq 0$ holds.

For instance, assume that we are dealing with real sectionally-continuous functions $x(u)$ in a fixed finite or infinite interval (a, b), assuming that in the case when the interval is infinite each of the functions $x(u)$ increases no more rapidly than a polynomial when the point u approaches the infinite end of the interval. The set of all such functions $x(u)$ may be taken to be the space E. The set of all those functions $x(u) \in E$ for which the inequality

$$x(u) \geqq 0 \quad (a < u < b),$$

holds forms the convex cone $\mathfrak{K}$. Finally we can take for the manifold $\mathfrak{M}$ the set of all real polynomials.‡

THEOREM 2.6.2. (CONCERNING THE EXTENSION OF A POSITIVE FUNCTIONAL). *For any $y \in E$, let the set of those $x \in \mathfrak{M}$ for which $x - y \in \mathfrak{K}$ be non-empty and the set of those $x \in \mathfrak{M}$ for which $y - x \in \mathfrak{K}$ likewise be non-empty. Then any $\mathfrak{K}$-positive functional $\mathfrak{P}\{x\}$ given in $\mathfrak{M}$ can be extended to the whole space E, with $\mathfrak{K}$-positivity conserved.*

Proof. We denote by y_1 any element of the space E which does not belong to $\mathfrak{M}$ and we introduce the manifold $\mathfrak{M}_1$ of elements

$$x_1 = x + \alpha y_1,$$

where x traverses $\mathfrak{M}$ and α traverses the set of all real numbers. We extend the functional $\mathfrak{P}$ to $\mathfrak{M}_1$ by putting

$$\mathfrak{K}\{x_1\} = \mathfrak{K}\{x\} + \alpha r_1.$$

No matter how we choose the number r_1 which evidently represents $\mathfrak{P}\{y_1\}$, the functional we have defined will be additive and homogeneous. Our problem consists in making such a choice of the number r_1 that the extended functional maintains the property of $\mathfrak{K}$-positivity. For this purpose we take the manifold $\mathfrak{U}_1^+$ of all those elements $x \in \mathfrak{M}$ for which $x - y_1 \in \mathfrak{K}$ (by assumption the set $\mathfrak{U}_1^+$ is not empty). Then we put

$$M_1 = \inf_{x \in \mathfrak{U}_1^+} \mathfrak{P}\{x\}.$$

† More precisely non-negative
‡ It is just for this concrete case that M. Riesz proves the possibility of the extension of a positive functional. However his construction can be taken over without any change for the case of an abstract space such as is the subject of Theorem 2.6.2. This construction can be found in later papers where it represents the basis of proof of other propositions, in particular of related ones.

Further we introduce the manifold $\mathfrak{N}_1^-$ (which is also non-empty) of those elements $x \in \mathfrak{M}$ for which $y_1 - x \in \mathfrak{K}$ and we put

$$m_1 = \sup_{x \in \mathfrak{N}_1^-} \mathcal{P}\{x\}.$$

By virtue of these definitions $m_1 > -\infty$ and $M_1 < \infty$. We now prove that

$$m_1 \leqq M_1.$$

At this point in our construction we use for the first time the convexity of the cone $\mathfrak{K}$. We take any elements $x'' \in \mathfrak{N}_1^+$ and $x' \in \mathfrak{N}_1^-$. This means that we have $x'' - y_1 \in \mathfrak{K}$ and $y_1 - x' \in \mathfrak{K}$; also, since the cone $\mathfrak{K}$ is convex the sum

$$(x'' - y_1) + (y_1 - x') = x'' - x'$$

also belongs to it. But we now apply the functional to the elements x', x'' and $x'' - x'$. Therefore

$$\mathcal{P}\{x''\} - \mathcal{P}\{x'\} = \mathcal{P}\{x'' - x'\} \geqq 0,$$

whence

$$\mathcal{P}\{x''\} \geqq \mathcal{P}\{x'\}.$$

Since this inequality will hold for any $x'' \in \mathfrak{N}_1^+$ and any $x' \in \mathfrak{N}_1^-$ we have that

$$\inf_{x'' \in \mathfrak{N}_1^+} \mathcal{P}\{x''\} \geqq \sup_{x' \in \mathfrak{N}_1^-} \mathcal{P}\{x'\},$$

i.e.

$$M_1 \geqq m_1.$$

For r_1 we may take any number which satisfies the inequality

$$m_1 \leqq r_1 \leqq M_1,$$

and is as easy to see, the functional $\mathcal{P}\{x\}$, which we obtain in $\mathfrak{M}_1$ will be $\mathfrak{K}$-positive.

The described construction is the first step; it is followed by analogous further steps, and with the aid of (possibly transfinite) induction the functional proves to be extended to the whole space E.

6.3. We now turn to the classical moment problem.

DEFINITION 2.6.3. The sequence $\{s_k\}_0^n$ $(n \leqq \infty)$ is called non-negative with respect to the interval (a, b), if for any polynomial $R(u)$ of degree $\leqq n$ the inequality

$$R(u) \geqq 0 \quad (a < u < b)$$

implies the inequality

$$S\{R\} \geqq 0.$$

Since the sequence $\{s_k\}_0^n$ is not necessarily assumed to be infinite, our considerations include in addition to the full moment problem also the truncated moment problem. The interval (a, b) may be either infinite or finite.† The essential part of the proof of the theorem to which we now pass does not depend on which of the cases just enumerated we are considering. However, to simplify the formulation we assume that in the case of an infinite interval (a, b) the right hand end is certainly infinite.

THEOREM 2.6.3.   (M. RIESZ[3])‡ *In order that a non-decreasing function $\sigma(u)$ ($a \leqq u \leqq b$) should exist for which*

$$s_k = \int_a^b u^k \, d\sigma(u) \quad (k = 0, 1, 2, ..., n-1) \qquad ...[2.47]$$

and

$$s_n \geqq \int_a^b u^n \, d\sigma(u), \qquad ...[2.48]$$

it is necessary and sufficient that the sequence $\{s_k\}_0^n$ should be non-negative with respect to the interval (a, b). In the case when the interval (a, b) is the whole number axis and $n < \infty$ this number n is assumed to be even.

If the interval (a, b) is finite, then the $\geqq$ sign in formula $[2.48]$ can be replaced by the equality sign.‖

The proof of necessity for the condition is trivial. In order to prove sufficiency we take any denumerable¶ point set $S = \{t_k\}_1^\infty$ which is everywhere dense in the interval (a, b). Further, we put

$$g_t(u) = \begin{cases} 1 & (a \leqq u < t), \\ 0 & (t \leqq u \leqq b). \end{cases}$$

The aggregate of all real polynomials of degree $\leqq n$, which we shall consider only in the interval (a, b), will be called $\mathfrak{M}$ and the linear space

---

† One could also replace the single interval (a, b) by a point set consisting of any system of intervals.

‡ M. Riesz discusses only the case when $n = \infty$ and the interval (a, b) is the whole number axis.

‖ We note that in the case $b < \infty$, $a = -\infty$, which we agreed not to consider, inequality $[2.48]$ should be replaced by

$$(-1)^n s_n \geqq \int_{-\infty}^b (-u)^n \, d\sigma(u).$$

¶ Because of this assumption we shall require only that case of Theorem 2.6.2, in which the complete extension of the functional is achieved without using transfinite induction.

obtained by adjoining to $\mathfrak{M}$ all functions $g_t(u)$ $(t \in S)$ will be called E. In this case the cone $\mathfrak{K}$ is the aggregate of all functions $x(u) \in E$ for which

$$x(u) \geqq 0 \quad (a < u < b).$$

The given positive sequence $\{s_k\}_0^n$ generates a $\mathfrak{K}$-positive functional S defined on the manifold $\mathfrak{M}$. With the aid of Theorem 2.6.2 we extend this functional to E and put

$$\mathsf{S}_u\{g_t(u)\} = \sigma(t).$$

In this way we obtain a certain function $\sigma(t)$ defined on S. We prove that this function is monotonic in the sense that

$$t'' > t', \; t' \in S, \; t'' \in S$$

implies

$$\sigma(t'') \geqq \sigma(t').$$

Indeed, by virtue of the additivity and homogeneity of the extended functional we can write down the equation

$$\sigma(t'') - \sigma(t') = \mathsf{S}_u\{g_{t''}(u)\} - \mathsf{S}_u\{g_{t'}(u)\} = \mathsf{S}_u\{g_{t''}(u) - g_{t'}(u)\}.$$

Also, since

$$g_{t''}(u) - g_{t'}(u) = \begin{cases} 0 & (a \leqq u < t'), \\ 1 & (t' \leqq u < t''), \\ 0 & (t'' \leqq u \leqq b) \end{cases}$$

and therefore $g_{t''}(u) - g_{t'}(u) \in \mathfrak{K}$, we have, by virtue of the $\mathfrak{K}$-positivity of the extended functional, that

$$\sigma(t'') - \sigma(t') \geqq 0.$$

As a result of the monotonic property of $\sigma(t)$ and the fact that the set S is dense in the interval (a, b) the function $\sigma(t)$ can be extended essentially uniquely to the whole interval (a, b), with its monotonic character preserved. Having performed this extension it remains to prove the truth of [2.47] and [2.48]. We shall perform this proof assuming for definiteness that $a = 0$, $b = \infty$ and $n < \infty$. In addition we may evidently assume that $0 \in S$. We choose points

$$0 = \tau_0 < \tau_{1_i} < \ldots < \tau_N = B \quad (B > 1, \tau_i = t_{k_i} \in S)$$

in such a way that the oscillation of the functions u^k $(0 \leqq k \leqq n-1)$ in each interval $[\tau_i, \tau_{i+1}]$ does not exceed a given number. Further we form the function

$$F_N^k(u) = \sum_{i=0}^{N-1} \tau_i^k\{g_{\tau_{i+1}}(u) - g_{\tau_i}(u)\}.$$

It is not difficult to verify that in the whole interval $[0, \infty)$ the following inequality holds:

$$0 \leqq u^k - F_N^k(u) \leqq \varepsilon + \frac{u^n}{B}. \qquad [2.49]$$

We apply the functional S to all terms of this inequality. We then find that

$$0 \leqq s_k - \sum_{i=0}^{N-1} \tau_i^k \{\sigma(\tau_{i+1}) - \sigma(\tau_i)\} \leqq \varepsilon s_0 + \frac{s_n}{B}.$$

For fixed B and $N \to \infty$ we obtain in the limit the inequality

$$0 \leqq s_k - \int_0^B t^k \, d\sigma(t) \leqq \frac{s_n}{B}.$$

It now remains to perform the limiting process $B \to \infty$, as a result of which we find precisely that

$$s_k = \int_0^\infty t^k \, d\sigma(t) \quad (k = 0, 1, 2, \ldots, n-1).$$

If we had taken $k = n$, then in place of the inequality $[2.49]$ we would have been able to write down the inequality

$$0 \leqq u^n - F_N^n(u) \quad (0 \leqq u < \infty),$$

whence we would have obtained that

$$0 \leqq s_n - \sum_{i=0}^{N-1} \tau_i^n \{\sigma(\tau_{i+1}) - \sigma(\tau_i)\}$$

and therefore

$$0 \leqq s_n - \int_0^B t^n \, d\sigma(t).$$

After passage to the limit $B \to \infty$ this gives us the inequality

$$s_n \geqq \int_0^\infty t^n \, d\sigma(t).$$

6.4. The power moment problem for the case of a finite interval (a, b) is called HAUSDORFF'S[2] moment problem. We shall dwell in a little more detail on this problem and for simplicity we shall take $a = 0$, $b = 1$. If we consider the truncated problem then we should extend Theorem 2.6.3 very slightly, namely by a criterion that the sequence $\{s_k\}_0^n$ should be non-negative with respect to the interval

[0, 1]. This criterion is easy to obtain from the well known theorem that states that any polynomial $R_n(u)$ of degree n which is $\geqq 0$ at all points of the interval [0, 1] can be represented in the form†

$$R_n(u) = u[A_m(u)]^2 + (1-u)[B_m(u)]^2,$$

if $n = 2m+1$ is odd, and in the form

$$R_n(u) = [C_m(u)]^2 + u(1-u)[D_{m-1}(u)]^2,$$

if $n = 2m$ is even. Here $A_m(u)$, $B_m(u)$, $C_m(u)$ and $D_{m-1}(u)$ are real polynomials, the degrees of which are given by their suffixes. If we use these representations we can formulate the following result: *the moment problem*

$$s_k = \int_0^1 u^k \, d\sigma(u) \quad (k = 0, 1, 2, \ldots, n; \ n < \infty)$$

has a non-decreasing solution if and only if the forms

$$\sum_{i, k=0}^{m} s_{i+k+1} x_i x_k, \quad \sum_{i, k=0}^{m} (s_{i+k} - s_{i+k+1}) x_i x_k \qquad \ldots [2.50]$$

for $n = 2m+1$ and the forms

$$\sum_{i, k=0}^{m} s_{i+k} x_i x_k, \quad \sum_{i, k=0}^{m-1} (s_{i+k+1} - s_{i+k+2}) x_i x_k \qquad \ldots [2.51]$$

for $n = 2m$ are non-negative.

The non-negative character of both forms in [2.50] or in [2.51] for all natural m is the necessary and sufficient condition for Hausdorff's moment problem to be soluble. However, in the case of the full moment problem there exists another criterion for solubility which has a number of advantages.

THEOREM **2.6.4.** (HAUSDORFF[2]) *For the moment problem*

$$s_k = \int_0^1 u^k \, d\sigma(u) \quad (k = 0, 1, 2, \ldots)$$

to be soluble within the class of non-decreasing functions it is necessary and sufficient that the inequalities

$$\Delta^m s_k \equiv \sum_{i=0}^{m} (-1)^i \binom{m}{i} s_{i+k} \geqq 0 \quad (m, k = 0, 1, 2, \ldots) \qquad \ldots [2.52]$$

should hold, which can be represented in the form

$$\leqq \{u^k(1-u)^m\} \geqq 0 \quad (m, k = 0, 1, 2, \ldots). \qquad \ldots [2.52a]$$

† The proof of this fact is given in Section 6.5.

The solution of this moment problem is unique.

Proof. The fact that the condition is necessary is trivial since

$$u^k(1-u)^m \geqq 0 \quad (0 \leqq u \leqq 1).$$

We now study the proof of sufficiency. Its essential part is the derivation of the identity

$$\sum_{k=0}^{N} \binom{N}{k} R\left(\frac{k}{N}\right) u^k (1-u)^{N-k} = R(u) + \sum_{i=1}^{n-1} \frac{E_i(u)}{N^i}, \quad \dots [2.53]$$

where $R(u)$ is an arbitrary polynomial of degree n, the $E_i(u)$ are polynomials of degrees $\leqq n$ in u which are independent of N, and $N \geqq n$. The left hand side of identity [2.53] is the so-called Bernstein polynomial for $R(u)$.

Let us assume for a moment that this identity has been established. We must prove that the inequality [2.52a] implies the non-negative property of the functional S. Assume that $R(u)$ is an arbitrary polynomial of degree n for which

$$R(u) \geqq 0 \quad (0 \leqq u \leqq 1).$$

Taking $N \geqq n$ and applying the functional S to both sides of [2.53] we find that

$$\mathsf{S}\{R(u)\} = \sum_{k=0}^{N} \binom{N}{k} R\left(\frac{k}{N}\right) \mathsf{S}\{u^k(1-u)^{N-k}\}$$
$$- \sum_{i=1}^{n-1} \frac{1}{N^i} \mathsf{S}\{E_i(u)\} \geqq - \sum_{i=1}^{n-1} \frac{1}{N^i} \mathsf{S}\{E_i(u)\}.$$

Hence, using the limiting process $N \to \infty$ we find that

$$\mathsf{S}\{R(u)\} \geqq 0.$$

This concludes the proof.

It remains for us to derive identity [2.53]. To do this we let the operator $\dfrac{1}{N} p \dfrac{d}{dp}$ act n times on the relation

$$\sum_{k=0}^{N} \binom{N}{k} p^k q^{N-k} = (p+q)^N,$$

where $N \geqq n$ and then put $p = u$, $q = 1-u$ in the resulting formula. On the left we obtain the polynomial

$$\sum_{k=0}^{N} \binom{N}{k} \left(\frac{k}{N}\right)^n u^k (1-u)^{N-k},$$

while on the right there will be a certain polynomial in u, the degree of which is evidently n. Thus we have the equation

$$\sum_{k=0}^{N} \binom{N}{k}\left(\frac{k}{N}\right)^n u^k (1-u)^{N-k} = u^n + \varepsilon_N(u),$$

where $\varepsilon_N(u)$ is a polynomial of degree $\leq n$ in u, the coefficients of which are rational fractions of N with the denominator N^{n-1}. We have on the left hand side the Bernstein polynomial of degree N for the function u^n. For $N \to \infty$ it tends to u^n uniformly in the interval $[0, 1]$. Therefore the remainder $\varepsilon_N(u)$ must have the form

$$\varepsilon_N(u) = \sum_{i=1}^{n-1} \frac{e_i(u)}{N^i},$$

where the $e_i(u)$ are polynomials of degree $\leq n$ in u which do not depend on N. Thus the required identity is proved for $R(u) = u^n$. But then it holds for any polynomial $R(u)$ of degree $\leq n$.

The uniqueness of the solution of this problem follows from the fact that in the case of a finite interval the polynomials form a dense set in L^2_σ for any distribution function $\sigma(u)$.

The proof we have expounded for the fact that a sequence $\{s_k\}_0^\infty$ which satisfies conditions [2.52] is non-negative, is due to HILDEBRANDT and SCHOENBERG[1]. Its advantage consists in the fact that it can be taken over directly to the many-dimensional case. Therefore 2.6.4 can also be generalized to the many-dimensional case† which was in fact proved by these authors.

6.5. *The full Stieltjes moment problem*

$$s_k = \int_0^\infty u^k \, d\sigma(u) \quad (k = 0, 1, 2, \ldots)$$

has a non-decreasing solution if and only if the two forms

$$\sum_{i,k=0}^{n} s_{i+k} x_i x_k, \quad \sum_{i,k=0}^{n} s_{i+k+1} x_i x_k$$

are non-negative for any n.

This proposition is a direct consequence of the general theorem **2.6.3** and the theorem which says that any polynomial $R(u)$ which satisfies the inequality

$$R(u) \geq 0 \quad (0 \leq u < \infty),$$

† See Addenda and Problems to Chapter 5, **15** where the theorem is formulated for the two-dimensional case.

can be represented in the form

$$R(u) = [A(u)]^2 + u[B(u)]^2, \qquad \qquad ...[2.54]$$

where $A(u)$ and $B(u)$ are real polynomials.

In the previous section we used a representation for a polynomial $T(x)$ which is non-negative in the interval $[0, 1]$. It is not difficult to see that this representation leads to the representation $[2.54]$ for the polynomial $R(u)$ which is non negative on the semi-axis $[0, \infty)$, and conversely. This is shown by means of the change of variable

$$x = \frac{u}{1+u}, \quad u = \frac{x}{1-x} \quad (0 \leqq u \leqq \infty, \quad 0 \leqq x \leqq 1)$$

and the relations

$$T(x) = \frac{R(u)}{(1+u)^n}, \quad R(u) = \frac{T(x)}{(1-x)^n},$$

where n is the degree of the polynomials $R(u)$ and $T(x)$.

It is appropriate to give here a direct proof of the representation $[2.54]$. We can assume that the leading coefficient of $R(u)$ is 1.

If $R(u)$ is a polynomial of the first degree we have

$$R(u) = u + R(0),$$

where $R(0) \geqq 0$, and therefore in this case the representation $[2.54]$ is obvious. Assume that $R(u)$ is a polynomial of the second degree

$$R(u) = u^2 + pu + q. \qquad \qquad ...[2.55]$$

It follows from $[2.55]$ that $q = R(0) \geqq 0$ and $R(\sqrt{q}) = \sqrt{q}(2\sqrt{q} + p) \geqq 0$. But in this case the equation

$$u^2 + pu + q = (u - \sqrt{q})^2 + (2\sqrt{q} + p)u$$

shows that the representation $[2.54]$ is true for a polynomial of the second degree. In order to use induction it remains to be noted that the equalities

$$R_1(u) = [A_1(u)]^2 + u[B_1(u)]^2$$

and

$$R_2(u) = [A_2(u)]^2 + u[B_2(u)]^2$$

imply the equality

$$R_1(u)R_2(u)$$

$$= [A_1(u)A_2(u) + uB_1(u)B_2(u)]^2 + u[A_1(u)B_2(u) - A_2(u)B_1(u)]^2.$$

6.6. We now turn again to M. Riesz' constructions given in Section **6.3.** It is easy to see, and was noted by Riesz himself, that in these constructions the special form of the moment functions is of no importance and that only certain of their general properties are significant, which can be brought out directly without great difficulty. Thus one can obtain a number of more or less general propositions. We shall give here one proposition of this kind which we shall use below.†

THEOREM **2.6.6.** *Suppose that I denotes the interval $(-\infty, b]$, where $b < \infty$. A certain set G of real continuous functions $\phi(u)$ $(u \in I)$ is given and on the linear hull $\mathfrak{M}$ of this set there is given a positive functional $\mathcal{P}$. Assume that $\mathfrak{M}$ contains a function $\psi(u)$ such that in the first place*

$$\inf_I \psi(u) > 0$$

and that secondly for any function $\phi(u) \in G$

$$\lim_{u \to -\infty} \frac{\phi(u)}{\psi(u)}$$

exists and is finite. In such a case one can find such a constant $M \geq 0$ and such a non-decreasing function of bounded variation $\sigma(u)$ $(-\infty \leq u \leq b)$, that for any function $\phi(u) \in G$ we have

$$\mathcal{P}\{\phi\} = c_\phi M + \int_{-\infty}^{b} \phi(u) \, d\sigma(u),$$

where

$$c_\phi = \lim_{u \to -\infty} \frac{\phi(u)}{\psi(u)}.$$

The theorem is true also for $I = (-\infty, \infty)$ if in addition to the conditions already indicated the function $\psi(u)$ also satisfies

$$\lim_{u \to \infty} \frac{\phi(u)}{\psi(u)} = \lim_{u \to -\infty} \frac{\phi(u)}{\psi(u)} \quad (\phi(u) \in G).$$

The detailed proof of this need not be given. We note merely the differences of this construction from that in Section **6.3.**

Taking an arbitrary function $\phi(u) \in G$ we form the difference

$$\theta(u) = \phi(u) - c_\phi \psi(u), \qquad \qquad \ldots[2.56]$$

then chose, as above, points $\tau_i = t_{k_i}$ for which

$$-A = \tau_0 < \tau_1 < \ldots < \tau_N = b,$$

---

† A more general proposition can be found in the article by Krein (see AKHIEZER and KREIN[1], Article II).

and form the functions

$$\theta_N(u) = \sum_{i=0}^{N-1} \theta(\tau_i)\{g_{\tau_{i+1}}(u) - g_{\tau_i}(u)\},$$

$$\psi_N(u) = \sum_{i=0}^{N-1} \psi(\tau_i)\{g_{\tau_{i+1}}(u) - g_{\tau_i}(u)\}.$$

We can convince ourselves easily of the fact that in the whole interval $(-\infty, b]$ the inequalities

$$-\varepsilon\psi(u) - \delta\psi(u) \leqq \theta(u) - \theta_N(u) \leqq \varepsilon\psi(u) + \delta\psi(u)$$

and

$$-\delta\psi(u) \leqq \psi(u) - \psi_N(u)$$

hold, where $\varepsilon = o(1)$ for $A \to \infty$ and depends only on the choice of the point A while δ can be made arbitrarily small for fixed A by choice of the number N and the points τ_i. It follows from the inequalities stated that

$$P\{\theta\} = \int_{-\infty}^{b} \theta(u)\,d\sigma(u),$$

$$P\{\psi\} \geqq \int_{-\infty}^{b} \psi(u)\,d\sigma(u).$$

Therefore

$$P\{\psi\} - \int_{-\infty}^{b} \psi(u)\,d\sigma(u) \equiv M \geqq 0,$$

and by virtue of equation [2.56]

$$P\{\phi\} - \int_{-\infty}^{b} \phi(u)\,d\sigma(u) = c_\phi M.$$

Addenda and Problems

1.† Assume that $\mathfrak{U}_{2n}$ denotes the aggregate of all polynomials $R(\lambda)$ of degree $\leqq 2n$ which satisfy the inequality

$$R(u) \geqq 0 \quad (-\infty < u < \infty),$$

† The following items 1–9 reproduce a number of results taken from the article by M. Riesz[1].

and assume that

$$T_{2n}(z) = \sup_{R \in \mathfrak{N}_{2n}} \frac{|R(z)|}{S\{R\}},$$

where z is an arbitrary complex point.
One can then assert that[†]

$$T_{2n}(z) \leq \frac{1}{\rho_n(z)} = h_n(z, \bar{z}).$$

Prove that

$$T_{2n}(z) \geq \frac{1}{2\rho_n(z)}.$$

Hint. Take the polynomial

$$R_0(\lambda) = \{h_n(\lambda, \bar{z}) + \varepsilon h_n(\lambda, z)\}\{h_n(\lambda, z) + \bar{\varepsilon} h_n(\lambda, \bar{z})\},$$

where

$$\varepsilon = \begin{cases} \dfrac{|h_n(z, z)|}{h_n(z, z)}, & \text{if } h_n(z, z) \neq 0, \\ 1, & \text{if } h_n(z, z) = 0. \end{cases}$$

We find that

$$\frac{|R_0(z)|}{S\{R_0\}} = \tfrac{1}{2}\{h_n(z, \bar{z}) + |h_n(z, z)|\}.$$

According to a remark by M. RIESZ[3] Prawitz has proved that the equality

$$T_{2n}(z) = \tfrac{1}{2}\{h_n(z, \bar{z}) + |h_n(z, z)|\}$$

is true.

2. Verify that $\lambda h_n(\lambda, 0)$ is a real quasiorthogonal polynomial of degree $n+1$. Denoting its zeros by $\theta_0 = 0, \theta_1, \theta_2, \ldots, \theta_n$, prove that

$$h_n(x, y) = \sum_{k=0}^{n} \frac{h_n(x, \theta_k) h_n(y, \theta_k)}{h_n(\theta_k, \theta_k)}.$$

Hint: The first statement is easy to verify. From it it follows by use of the Darboux-Christoffel-formula that

$$\sum_{0}^{n} P_k(x) P_k(y) = A \frac{y P_n(x) h_n(y, 0) - x P_n(y) h_n(x, 0)}{x - y},$$

[†] Here and in the following $h_n(x, y)$ denotes the kernel polynomial belonging to the functional S.

where A is a constant, and therefore

$$h_n(\theta_i, \theta_j) = 0 \quad (i \neq j).$$

We then write down Lagrange's interpolation formula

$$h_n(x, y) = \sum_{k=0}^{n} h_n(x, \theta_k) \frac{\omega(y)}{\omega'(\theta_k)(y - \theta_k)},$$

where $\omega(y) = y h_n(y, 0)$. Putting $x = \theta_i$ we find that

$$h_n(y, \theta_i) = h_n(\theta_i, \theta_i) \frac{\omega(y)}{\omega'(\theta_i)(y - \theta_i)}.$$

3. Prove that

$$h_n(x, y) \begin{vmatrix} s_2 & s_3 & \cdots & s_{n+1} \\ s_3 & s_4 & \cdots & s_{n+2} \\ \cdot & \cdot & \cdot & \cdot \\ s_{n+1} & s_{n+2} & \cdots & s_{2n} \end{vmatrix}$$

$$= D_n h_n(x, 0) h_n(y, 0) - xy \begin{vmatrix} 0 & 1 & x & \cdots & x^{n-1} \\ 1 & s_2 & s_3 & \cdots & s_{n+1} \\ y & s_3 & s_4 & \cdots & s_{n+2} \\ \cdot & \cdot & \cdot & \cdot & \cdot \\ y^{n-1} & s_{n+1} & s_{n+2} & \cdots & s_{2n} \end{vmatrix}.$$

Hint. To the determinant

$$\begin{vmatrix} 0 & 1 & x & \cdots & x^n \\ 1 & s_0 & s_1 & \cdots & s_n \\ y & s_1 & s_2 & \cdots & s_{n+1} \\ \cdot & \cdot & \cdot & \cdot & \cdot \\ y^n & s_n & s_{n+1} & \cdots & s_{2n} \end{vmatrix}$$

apply the theorem concerning the minors of the adjugate determinant.

4. Assume that $\rho_n^{(2)}(z)$ and $\rho^{(2)}(z)$ are determined for the sequence $\{s_{k+2}\}_{k=0}^{\infty}$ in the same way as $\rho_n(z)$ and $\rho(z)$ are determined for the sequence $\{s_k\}_0^{\infty}$. Thus in particular

$$\frac{1}{\rho_n^{(2)}(z)} \begin{vmatrix} s_2 & s_3 & \cdots & s_{n+2} \\ s_3 & s_4 & \cdots & s_{n+3} \\ \cdot & \cdot & \cdot & \cdot \\ s_{n+2} & s_{n+3} & \cdots & s_{2n+2} \end{vmatrix} = - \begin{vmatrix} 0 & 1 & z & \cdots & z^n \\ 1 & s_2 & s_3 & \cdots & s_{n+2} \\ \bar{z} & s_3 & s_4 & \cdots & s_{n+3} \\ \cdot & \cdot & \cdot & \cdot & \cdot \\ \bar{z}^n & s_{n+2} & s_{n+3} & \cdots & s_{2n+2} \end{vmatrix}.$$

Prove that for real x

$$\frac{1}{\rho_n(x)} = \frac{1}{\rho_n(0)} \frac{[h_n(x, 0)]^2}{[h_n(0, 0)]^2} + \frac{x^2}{\rho_{n-1}^{(2)}(x)}.$$

Proof. Put $y = x$ in example **3**.

5. Using the results of examples **2, 3** and **4** and the hint to example **2** prove that for real x

$$\frac{1}{\rho_{n-1}^{(2)}(x)} = \sum_{k=1}^{n} \left[\frac{h_n(x, 0)}{\theta_k h_n'(\theta_k, 0)(x - \theta_k)} \right]^2 \frac{1}{\rho_n(\theta_k)}.$$

Hence obtain the inequality

$$\frac{1}{\rho_{n-1}^{(2)}(x)} \leqq \frac{1}{\rho_{n-1}^{(2)}(0)} \max_{k = 1, 2, \ldots, n} \left[\frac{h_n(x, 0)}{h_n(0, 0)\left(1 - \dfrac{x}{\theta_k}\right)} \right]^2,$$

which is valid for any real x.

6. Prove that

$$\left| \sum_{1}^{n} \frac{1}{\theta_k} \right| \leqq \sqrt{\frac{s_0 - \rho_n(0)}{\rho_{n-1}^{(2)}(0)}},$$

$$\sum_{1}^{n} \frac{1}{\theta_k^2} \leqq 4\frac{s_0 - \rho_n(0)}{\rho_{n-1}^{(2)}(0)}.$$

Hint. Introduce polynomials of degree n

$$K(u) = 1 - \frac{h_n(u, 0)}{h_n(0, 0)}, \quad L(u) = \frac{1}{u}K(u) + K'(0)[K(u) - 1].$$

It is easy to see that

$$K(\theta_i) = 1 \quad (i = 1, 2, \ldots, n), \quad K(0) = 0, \quad K'(0) = \sum_{1}^{n} \frac{1}{\theta_k},$$

$$L(\theta_i) = \frac{1}{\theta_i} \quad (i = 1, 2, \ldots, n), \quad L(0) = 0,$$

$$L'(0) = \frac{1}{2}\left(\sum_{1}^{n} \frac{1}{\theta_k} \right)^2 + \frac{1}{2}\sum_{1}^{n} \frac{1}{\theta_k^2}.$$

We have the equation

$$s_0 = \sum_{k=0}^{n} \rho_n(\theta_k),$$

whence

$$s_0 - \rho_n(0) = \sum_{k=1}^{n} \rho_n(\theta_k) = \sum_{i=0}^{n} [K(\theta_i)]^2 \rho_n(\theta_i) = \mathfrak{S}\{[K(u)]^2\}$$

$$= \int_{-\infty}^{\infty} \left[\frac{K(u)}{u}\right]^2 u^2 \, d\sigma(u) \geqq [K'(0)]^2 \rho_{n-1}^{(2)}(0)$$

and therefore

$$|K'(0)| \leqq \sqrt{\frac{s_0 - \rho_n(0)}{\rho_{n-1}^{(2)}(0)}}.$$

Analogously, considering the polynomial $L(u)$ we can prove the second inequality.

7. Prove that for any real x

$$\frac{1}{\rho(x)} \leqq \left\{\frac{1}{\rho(0)} + \frac{x^2}{\rho^{(2)}(0)}\right\} \exp\{4A^2x^2 + 6A \mid x \mid\},$$

where

$$A = \sqrt{\frac{s_0 - \rho(0)}{\rho^{(2)}(0)}}.$$

Hint. Use example **6** and prove that

$$\left[\frac{h_n(x, 0)}{h_n(0, 0)}\right]^2 \leqq \exp\{4A_n^2 x^2 + 2A_n \mid x \mid\};$$

$$\left[\frac{h_n(x, 0)}{h_n(0, 0)\left(1 - \dfrac{x}{\theta_k}\right)}\right]^2 \leqq \exp\{4A_n^2 x^2 + 6A_n \mid x \mid\},$$

where

$$A_n = \sqrt{\frac{s_0 - \rho_n(0)}{\rho_{n-1}^{(2)}(0)}}.$$

8. For the moment problem

$$s_k = \int_{-\infty}^{\infty} u^k \, d\sigma(u) \quad (k = 0, 1, 2, \ldots)$$

to be determinate it is necessary and sufficient that

$$\rho(0)\rho^{(2)}(0) = 0.$$

Proof. If the moment problem is indeterminate then $\rho(x) \neq 0$ $(-\infty < x < \infty)$, therefore by virtue of example **4**, $\rho^{(2)}(x) \neq 0$ $(-\infty < x < \infty)$ and so

$$\rho(0)\rho^{(2)}(0) \neq 0.$$

Conversely if

$$\rho(0)\rho^{(2)}(0) \neq 0,$$

then by virtue of example **7** $\rho(x) \neq 0$ $(-\infty < x < \infty)$ and therefore the moment problem is indeterminate (see Corollary **2.5.3**).

9. From the theorem of example **8** obtain HAMBURGER's[3] criterion: For a moment problem

$$s_k = \int_{-\infty}^{\infty} u^k \, d\sigma(u) \quad (k = 0, 1, 2, \ldots)$$

to be determinate it is necessary and sufficient that

$$\lim_{n \to \infty} \frac{\begin{vmatrix} s_0 & s_1 & \cdots & s_n \\ s_1 & s_2 & \cdots & s_{n+1} \\ \cdot & \cdot & \cdots & \cdot \\ s_n & s_{n+1} & \cdots & s_{2n} \end{vmatrix}}{\begin{vmatrix} s_4 & s_5 & \cdots & s_{n+2} \\ s_5 & s_6 & \cdots & s_{n+3} \\ \cdot & \cdot & \cdots & \cdot \\ s_{n+2} & s_{n+3} & \cdots & s_{2n} \end{vmatrix}} = 0.$$

10. For the moment problem

$$s_k = \int_{-\infty}^{\infty} u^k \, d\sigma(u) \quad (k = 0, 1, 2, \ldots) \qquad \ldots[2.57]$$

to be determinate it is necessary and sufficient that at least one of the series

$$\sum_0^{\infty} [P_k(0)]^2, \quad \sum_1^{\infty} [Q_k(0)]^2 \qquad \ldots[2.58]$$

be divergent (HAMBURGER[3]).

The proof of sufficiency is trivial. In order to prove the necessary condition we assume that both series $[2.58]$ converge. From the convergence of the first series it follows that the moment problem has a solution with the mass

$$\rho(0) = \frac{1}{\sum_0^{\infty} [P_k(0)]^2}$$

at the point $u = 0$. On the other hand, owing to the convergence of the second series we can consider the sequence s_{-2}, s_{-1}; $s_0, s_1, \ldots$ where

$$s_{-2} = \sum_1^\infty [Q_k(0)]^2,$$

$$s_{-1} = 0.$$

This new sequence is positive. Indeed (see Addenda and Problems to Chapter 1 example 6)

$$\begin{vmatrix} s_{-2} & s_{-1} & s_0 & \cdots & s_{n-1} \\ s_{-1} & s_0 & s_1 & \cdots & s_n \\ \cdots & \cdots & \cdots & \cdots & \cdots \\ s_{n-1} & s_n & s_{n+1} & \cdots & s_{2n} \end{vmatrix} = s_{-2}D_n + \begin{vmatrix} 0 & 0 & s_0 & \cdots & s_{n-1} \\ 0 & s_0 & s_1 & \cdots & s_n \\ \cdots & \cdots & \cdots & \cdots & \cdots \\ s_{n-1} & s_n & s_{n+1} & \cdots & s_{2n} \end{vmatrix}$$

$$= D_n \sum_{n+1}^\infty [Q_k(0)]^2 > 0.$$

Therefore there exists a non-decreasing function $\tau(u)$ for which

$$s_m = \int_{-\infty}^\infty u^m \, d\tau(u) \quad (m = -2, -1, 0, 1, \ldots).$$

Thus the moment problem [2.57] has the solution

$$\sigma^*(u) = \int_{-\infty}^u t^2 \, d\tau(t),$$

which has zero mass at the point $u = 0$. This solution is thus different from that previously given and therefore the moment problem is indeterminate.

11. If

$$\sum_1^\infty \frac{1}{\sqrt[2n]{s_{2n}}} = \infty,$$

the moment problem

$$s_k = \int_{-\infty}^\infty u^k \, d\sigma(u) \quad (k = 0, 1, 2, \ldots)$$

is determinate (CARLEMAN[3]).

Proof. From the equation

$$b_{k-1}y_{k-1} + a_k y_k + b_k y_{k+1} = \lambda y_k$$

it follows that

$$b_0 b_1 \ldots b_{n-1} P_n(\lambda) = \lambda^n + \ldots .$$

Hence

$$b_0 b_1 \ldots b_{n-1} \int_{-\infty}^{\infty} [P_n(u)]^2 \, d\sigma(u) = \int_{-\infty}^{\infty} u^n P_n(u) \, d\sigma(u)$$

$$\leq \sqrt{\int_{-\infty}^{\infty} u^{2n} \, d\sigma(u)} \sqrt{\int_{-\infty}^{\infty} [P_n(u)]^2 \, d\sigma(u)}$$

and therefore

$$b_0 b_1 \ldots b_{n-1} \leq \sqrt{s_{2n}}.$$

Therefore

$$\sum_1^\infty \frac{1}{\sqrt[2n]{s_{2n}}} \leq \sum_1^\infty \frac{1}{\sqrt[n]{b_0 b_1 \ldots b_{n-1}}}.$$

Now it remains to use Carleman's criterion (see Addenda and Problems to Chapter 1, example 1) and also CARLEMAN'S[3] general inequality†

$$\sum_1^\infty \sqrt[n]{u_1 u_2 \ldots u_n} < e \sum_1^\infty u_k,$$

where the u_k are non-negative real numbers not all of which are zero.

12. If $\phi(u) > 0$ $(-\infty < u < \infty)$ is a measureable function and if for certain values $q \geq 1, \delta > 0$ we have

$$\int_{-\infty}^{\infty} e^{\delta |u|} \frac{du}{[\phi(u)]^q} < \infty,$$

† We give a proof of Carleman's inequality due to POLYA[1].
We define numbers $c_1, c_2, c_3, \ldots$ by the equations

$$c_1 c_2 \ldots c_n = (n+1)^n,$$

so that

$$c_n = \frac{(n+1)^n}{n^{n-1}} < ne,$$

and we use the inequality between arithmetic and geometric means:

$$\sum_1^\infty (u_1 u_2 \ldots u_n)^{\frac{1}{n}} = \sum_1^\infty \frac{(u_1 c_1 u_2 c_2 \ldots u_n c_n)^{\frac{1}{n}}}{n+1}$$

$$\leq \sum_1^\infty \frac{u_1 c_1 + u_2 c_2 + \ldots + u_n c_n}{n(n+1)} = \sum_{k=1}^\infty u_k c_k \sum_{n=k}^\infty \frac{1}{n(n+1)}$$

$$= \sum_{k=1}^\infty u_k c_k \sum_{n=k}^\infty \left(\frac{1}{n} - \frac{1}{n+1} \right) = \sum_{k=1}^\infty u_k c_k \frac{1}{k} < e \sum_{k=1}^\infty u_k.$$

then the sequence

$$s_k = \int_{-\infty}^{\infty} u^k \frac{du}{\phi(u)} \quad (k = 0, 1, 2, \ldots)$$

generates a determinate moment problem (HARDY[1])

Hint. Bound the moments s_{2n} from above and apply Carleman's criterion (example **11**).

The case $q > 1$ reduces to the case $q = 1$ with the aid of Hölder's inequality. For $q = 1$ we have

$$s_{2n} = \int_{-\infty}^{\infty} u^{2n} e^{-\delta |u|} e^{\delta |u|} \frac{du}{\phi(u)} \leq \max_t \{ t^{2n} e^{-\delta |t|} \} \int_{-\infty}^{\infty} e^{\delta |u|} \frac{du}{\phi(u)}$$

$$= (2n)^{2n} (\delta e)^{-2n} \int_{-\infty}^{\infty} e^{\delta |u|} \frac{du}{\phi(u)}.$$

13. Assume that $\sigma(u)$ $(-\infty < u < \infty)$ is a non-decreasing function of bounded variation. In this case, in order that the linear hull of the functions $e^{i\alpha u}$ be dense in L_σ^p $(p \geq 1)$, provided α takes on only positive values and 0, it is necessary and sufficient that

$$\int_{-\infty}^{\infty} \frac{\ln \sigma'(u)}{1 + u^2} du = -\infty,$$

where $\sigma'(u)$ is the derivative of the absolutely continuous part of the function $\sigma(u)$ (KREIN[5] for $p = 2$, AKHIEZER[2] for arbitrary $p \geq 1$).

14. Assume that $\sigma(u)$ $(-\infty < u < \infty)$ is a non-decreasing function which possesses finite moments of all orders

$$s_k = \int_{-\infty}^{\infty} u^k \, d\sigma(u) \quad (k = 0, 1, 2, \ldots).$$

If

$$\int_{-\infty}^{\infty} \frac{\ln \sigma'(u)}{1 + u^2} du > -\infty,$$

then the sequence $\{s_k\}_0^\infty$ generates an indeterminate moment problem (Krein).

Proof. By virtue of the conditions and the proposition of example **13** there exists a function $f(u) \in L_\sigma^2$ for which

$$\int_{-\infty}^{\infty} |f(u)|^2 \, d\sigma(u) \neq 0, \qquad \int_{-\infty}^{\infty} f(u) e^{i\alpha u} d\sigma(u) = 0 \quad (\alpha \geq 0).$$

Differentiating the last relation with respect to α and putting $\alpha = 0$, we find that

$$\int_{-\infty}^{\infty} f(u)\, u^k\, d\sigma(u) = 0 \quad (k = 0, 1, 2, ...),$$

whence it follows that the set of all polynomials is not dense in L_σ^2. Therefore we have an indefinite moment problem.

15. Each of the sequences given below generates an indeterminate Stieltjes moment problem:

a) $s_k = \int_0^\infty u^{-\ln u}\, u^k\, du$ (STIELTJES[3])

b) $s_k = \int_0^\infty e^{-\sqrt[4]{u}}\, u^k\, du$ (STIELTJES[3])

c) $s_k = \int_0^\infty \exp\left\{-\frac{\pi\sqrt{u}}{\ln^2 u + \pi^2}\right\} u^k\, du.$ (HAMBURGER[1])

Stieltjes' proof is based on the consideration of certain continued fractions or on the evaluation of certain integrals. For instance the proof in case a) follows from the fact that

$$\int_0^\infty u^{-\ln u}\, u^k \sin(2\pi \ln u)\, du = 0 \quad (k = 0, 1, 2, ...).$$

In a similar way in the case c) Hamburger proves that

$$\int_0^\infty \exp\left\{-\frac{\pi\sqrt{u}}{\ln^2 u + \pi^2}\right\} u^k g(u)\, du = 0 \quad (k = 0, 1, 2, ...),$$

where

$$g(u) = \sin\frac{\sqrt{u}\,\ln u + \pi}{\ln^2 u + \pi^2} \cdot \exp\left\{\frac{\ln u}{\ln^2 u + \pi^2}\right\}.$$

Show that all three results are consequences of the general theorem of example **14.**

16. Assume that $\phi(z)$ is an entire transcendental function of zero genus, the zeros of which all lie in a certain strip of finite width parallel to the imaginary axis and that

$$\phi(u) > 0 \quad (-\infty < u < \infty), \quad \phi(0) = 1.$$

In this case,
a) the integrals

$$s_k = \int_{-\infty}^{\infty} u^k \frac{du}{\phi(u)} \quad (k = 0, 1, 2, \ldots)$$

exist and generate an indeterminate moment problem;
b) if one puts

$$\sigma(u) = \int_{-\infty}^{u} \frac{dt}{\phi(t)}$$

and denotes by G the closure in L_σ^2 of the set of all polynomials then $L_\sigma^2 \ominus G$ coincides with the set of all functions $F(u)$ $(-\infty < u < \infty)$ of the form

$$F(u) = \frac{\omega(u)}{\sqrt{2\pi}} \int_0^\infty e^{itu} h(t) \, dt + \frac{\overline{\omega}(u)}{\sqrt{2\pi}} \int_{-\infty}^0 e^{itu} h(t) \, dt,$$

where $h(t) \in L^2(-\infty, \infty)$ and $\omega(u)$ and $\overline{\omega}(u)$ are determined by means of the expansion

$$\phi(z) = \omega(z) \overline{\omega}(z), \quad \omega(z) = \prod_{k=1}^{\infty} \left(1 - \frac{z}{c_k}\right) \quad (\text{Im } c_k > 0).$$

Here the inequality

$$\int_{-\infty}^{\infty} |F(u)|^2 \frac{du}{\phi(u)} = \int_{-\infty}^{\infty} |h(t)|^2 \, dt$$

holds (AKHIEZER[6])†.

† See also the note by GURARII[1] in which there is a generalisation of the authors' result.

Chapter 3

FUNCTION THEORETIC METHODS IN THE
MOMENT PROBLEM

The power moment problem can be included as a limiting case of a
general interpolation problem in the theory of functions the statement
and solution of which is associated with a number of prominent names.
(Carathéodory, Schur, Nevanlinna). Other cases of this interpolation
problem can also be considered as certain moment problems with the
difference that in them one takes in place of the powers u^k ($k = 0, 1, 2, ...$)
certain other " moment " functions.

The functional theoretical approach to the power moment problem
opens up ways for the discussion and solution of this problem. In
particular it led to the description of all solutions of this problem in the
indeterminate case. This was done for the first time by Nevanlinna.
The formula which he obtained was the source of a number of new
questions concerning the solution of the indeterminate problem.

1. An interpolation problem in the theory of analytic functions

1.1 The interpolation problem to which we refer consists of the
following: in the planes of the complex variables z and w regions G_z
and G_w are given; further, two point sets are given: $Z = \{z_\alpha\} \subset G_z$
and $\{w_a\} \subset G_w$; it is required to find a function $w = w(z)$ analytic in
$G(z)$ (which may not exist), the range of whose values belongs to G_w
and which satisfies the conditions

$$w(z_\alpha) = w_\alpha \quad (z_\alpha \in Z);$$

if the set Z contains multiple points these conditions must be modified
in a natural way; a modification is also necessary in the limiting case
when a point $z_\alpha \in Z$ moves to the boundary of the region G_z.

We shall assume that the regions G_z and G_w are simply connected.
In this case we can take these regions to be either half-planes or unit
circles. This is what is done usually. It then proves useful to single

out certain particular cases and to introduce special classes of analytic functions corresponding to these special cases.

a) The class C (CARATHÉODORY[1]) consists of all functions $w = F(z)$ when G_z is the circle $|z| < 1$ and G_w the half-plane Re $w \geqq 0$.

The class C is characterised by an integral representation (F. REISZ[1], HERGLOTZ[1])

$$F(z) = iv + \frac{1}{2\pi} \int_{-\pi}^{\pi} \frac{e^{i\theta} + z}{e^{i\theta} - z} \, d\sigma(\theta), \qquad \dots[3.1]$$

where

$$v = \text{Im } F(0),$$

and $\sigma(\theta)$ is a non-decreasing function of bounded variation which is essentially uniquely determined by the function $F(z)$. The representation [3.1] is obtained with the aid of Helly's theorems and the Schwartz formula which expresses an analytic function in a circle in terms of the values of its real part on the boundary. Indeed let

$$F(z) = u(r, \phi) + iv(r, \phi) \quad (z = re^{i\phi}).$$

Then we have for $|z| < R < 1$:

$$F(z) = i \text{ Im } F(0) + \frac{1}{2\pi} \int_{-\pi}^{\pi} \frac{Re^{i\theta} + z}{Re^{i\theta} - z} u(R, \theta) \, d\theta$$

and therefore

$$F(Rz) = i \text{ Im } F(0) + \frac{1}{2\pi} \int_{-\pi}^{\pi} \frac{e^{i\theta} + z}{e^{i\theta} - z} \, d\sigma_R(\theta) \quad (0 < R < 1, |z| < 1),$$

$$\dots[3.1a]$$

where $\sigma_R(\theta)$ is a non-decreasing function defined by the equation

$$\sigma_R(\theta) = \int_{-\pi}^{\theta} u(R, \psi) \, d\psi \quad (-\pi \leqq \theta \leqq \pi).$$

The family of functions $\sigma_R(\theta)$ $(0 < R < 1)$ is uniformly bounded, since

$$0 \leqq \sigma_R(\theta) \leqq \sigma_R(2\pi) = 2\pi \cdot \text{Re } F(0).$$

Therefore Helly's first theorem is applicable by virtue of which one can find a sequence $\{\sigma_{R_k}(\theta)\}_{k=1}^{\infty}$, $R_k \to 1$, and a non-decreasing function $\sigma(\theta)$ at all points of continuity of which one has

$$\lim_{k \to \infty} \sigma_{R_k}(\theta) = \sigma(\theta).$$

Replacing R by R_k in equation [3.1a] and passing to the limit we obtain the required representation [3.1] from Helly's second theorem.

b) The class S (SCHUR[2]) consists of all functions $w = \phi(z)$ when G_z is the circle $|z| < 1$, and G_w the circle $|w| \leqq 1$.

There exists a simple relation between the classes C and S; it is given by the formula

$$\phi(z) = \frac{1}{z}\frac{F(z)-F(0)}{F(z)+\overline{F(0)}}, \qquad \ldots[3.2]$$

where $F(z)$ ranges over class C and $\phi(z)$ over class S.

c) The class N (NEVANLINNA[1]) consists of all functions $w = f(z)$ when G_z is the half-plane Im $z > 0$ and G_w is the half-plane Im $w \geqq 0$.

Here one also has an integral representation, which is easily obtained from the Riesz-Herglotz formula, namely

$$f(z) = \mu z + v + \int_{-\infty}^{\infty} \frac{1+uz}{u-z}\, d\tau(u), \qquad \ldots[3.3]$$

where μ, v are real constants, the first of which is $\geqq 0$ while $\tau(u)$ is a non-decreasing function of bounded variation.† For brevity we shall agree to call equation [3.3] Nevanlinna's formula. We have already encountered a particular case of this formula, namely

$$f(z) = \int_{-\infty}^{\infty} \frac{d\sigma(u)}{u-z}, \qquad \ldots[3.3a]$$

where $\sigma(u)$ is a non-decreasing function of bounded variation. This formula is the particular case of formula [3.3] when

$$\int_{-\infty}^{\infty} (1+u^2)\, d\tau(u) < \infty.$$

Indeed, putting

$$\sigma(u) = \int_{-\infty}^{u} (1+v^2)\, d\tau(v),$$

we can rewrite [3.3] in the form

$$f(z) = \mu z + v + \int_{-\infty}^{\infty} \left(\frac{1}{u-z} - \frac{u}{1+u^2} \right) d\sigma(u)$$

† This variation is

$$\operatorname{Im} f(i) - \mu.$$

and we obtain equation [3.3a], if

$$\mu = 0, \quad v = \int_{-\infty}^{\infty} \frac{u}{1+u^2} \, d\sigma(u).$$

It is easy to prove that for $f(z)$ (Im $z > 0$) to be capable of representation in the form [3.3a] it is necessary and sufficient that $f(z) \in N$ and that the inequality

$$C = \sup_{y \geq 1} | yf(iy) | < \infty$$

should hold.

Only the proof of sufficiency is non-trivial. We now indicate this proof. If $f(z) \in N$, then the representation [3.3] holds. Therefore

$$yf(iy) = i\mu y^2 + vy + y \int_{-\infty}^{\infty} \frac{1+iuy}{u-iy} \, d\tau(u),$$

whence

$$\text{Im} \left[yf(iy) \right] = \mu y^2 + y^2 \int_{-\infty}^{\infty} \frac{1+u^2}{u^2+y^2} \, d\tau(u),$$

$$\text{Re} \left[yf(iy) \right] = vy + y \int_{-\infty}^{\infty} \frac{u(1-y^2)}{u^2+y^2} \, d\tau(u).$$

By virtue of our condition we have

$$\text{Im} \left[yf(iy) \right] \leq C \quad (y \geq 1).$$

Therefore $\mu = 0$ and for any $A > 0, B > 0$ we have

$$y^2 \int_{-A}^{B} \frac{1+u^2}{u^2+y^2} \, d\tau(u) \leq C,$$

whence it follows that

$$\int_{-A}^{B} (1+u^2) \, d\tau(u) \leq C$$

and therefore

$$\int_{-\infty}^{\infty} (1+u^2) \, d\tau(u) \leq C.$$

By virtue of our condition the following inequality also holds:

$$| \text{Re} \left[yf(iy) \right] | \leq C \quad (y \geq 1).$$

Therefore

$$v + \lim_{y \to \infty} \int_{-\infty}^{\infty} \frac{u(1-y^2)}{u^2+y^2} \, d\tau(u) = 0,$$

whence

$$v = \int_{-\infty}^{\infty} u \, d\tau(u).$$

On the basis of this equation and the equation $\mu = 0$ the representation [3.3] takes on the form

$$f(z) = \int_{-\infty}^{\infty} \frac{(1+u^2) \, d\tau(u)}{u-z} = \int_{-\infty}^{\infty} \frac{d\sigma(u)}{u-z}.$$

Any function $f(z) \in N$ can be continued on to the half-plane $\operatorname{Im} z < 0$ by putting

$$f(z) = \overline{f(\bar{z})} \quad (\operatorname{Im} z < 0).$$

Of course the functions obtained in this way in the half planes $\operatorname{Im} z > 0$ and $\operatorname{Im} z < 0$ are not in general analytic continuations of one another. However, for many purposes such an extension is useful and we shall sometimes employ it. Thus we can understand the class N to consist of all functions $f(z)$ that are regular for $\operatorname{Im} z > 0$ and $\operatorname{Im} z < 0$ and which satisfy the relation

$$f(z) = \overline{f(\bar{z})}, \quad \frac{\operatorname{Im} f(z)}{\operatorname{Im} z} \geqq 0 \quad (\operatorname{Im} z \neq 0).$$

The integral representation of functions of this class is equation [3.3]. We stress that the constants involved in this representation are real.†

The classes of functions which we have introduced in this section do not possess generally accepted names. For instance the class C is sometimes called the class of pseudo-positive functions and the class N is often called the class R (to stress its relation to the so-called Riesz resolvents of self-adjoint operators).

† The function

$$f(z) = \beta + i\gamma \quad (\operatorname{Im} z > 0),$$

where $\beta \gtrless 0$ and $\gamma > 0$ are constants, if included in the class N, should not be considered as a constant. Its integral representation is

$$\beta + \frac{\gamma}{\pi} \int_{-\infty}^{\infty} \frac{1+uz}{u-z} \frac{du}{1+u^2}.$$

2. Reduction of the power moment problem to a certain problem in the theory of functions

2.1. The problem in the theory of functions which will be discussed consists in finding functions of the class N which have a given asymptotic representation.

THEOREM **3.2.1.** (HAMBURGER[3], NEVANLINNA[1]) *If a non-decreasing function $\sigma(u)$ $(-\infty < u < \infty)$ possesses finite moments*

$$s_k = \int_{-\infty}^{\infty} u^k \, d\sigma(u) \quad (k = 0, 1, 2, ..., 2n), \qquad ...[3.4]$$

then there exists a function $f(z)$ in the class N, namely

$$f(z) = \int_{-\infty}^{\infty} \frac{d\sigma(u)}{u-z}, \qquad ...[3.5]$$

such that given any fixed $\delta > 0 (< \frac{1}{2}\pi)$ however small, it is true uniformly in the range of angles

$$\delta \leq \arg z \leq \pi - \delta \qquad ...[3.6]$$

that

$$\lim_{z \to \infty} z^{2n+1} \left\{ f(z) + \frac{s_0}{z} + \frac{s_1}{z^2} + ... + \frac{s_{2n-1}}{z^{2n}} \right\} = -s_{2n}. \quad ...[3.7]$$

Conversely if for some function $f(z) \in N$ the relation [3.7] holds with real numbers s_k, at least for $z = iy$ ($y \to \infty$), then the function $f(z)$ permits the representation [3.5] where $\sigma(u)$ is a non-decreasing function whose moments are [3.4].

To prove the first statement we write down the following equation which follows from [3.5] and [3.4]

$$z^{2n+1} \left\{ f(z) + \frac{s_0}{z} + \frac{s_1}{z^2} + ... + \frac{s_{2n-1}}{z^{2n}} \right\}$$

$$= \int_{-\infty}^{\infty} \frac{zu^{2n}}{u-z} \, d\sigma(u) = -s_{2n} + \int_{-\infty}^{\infty} \frac{u^{2n+1}}{u-z} \, d\sigma(u).$$

Thus it has to be proved that the integral on the right hand side tends uniformly to zero for $z \to \infty$ within the angle [3.6]. But if z lies in that angle then we have for any real u:

$$|z-u| \geq |z| \sin \delta, \quad |z-u| \geq |u| \sin \delta.$$

Therefore

$$\left| \int_{-\infty}^{\infty} \frac{u^{2n+1}}{u-z} \, d\sigma(u) \right| \leqq \frac{1}{|z| \sin \delta} \int_{-A}^{A} |u|^{2n+1} \, d\sigma(u)$$

$$+ \frac{1}{\sin \delta} \int_{|u| > A} u^{2n} \, d\sigma(u),$$

which proves the first statement.

We prove the second statement. We have the equation

$$\lim_{y \to \infty} (iy)^{2n+1} \left\{ f(iy) + \frac{s_0}{iy} + \frac{s_1}{(iy)^2} + \ldots + \frac{s_{2n-1}}{(iy)^{2n}} \right\} = -s_{2n}. \quad \ldots [3.8]$$

Hence in particular

$$\lim_{y \to \infty} iyf(iy) = -s_0 \qquad \ldots [3.8a]$$

and therefore

$$\sup_{y \geqq 1} |yf(iy)| < \infty.$$

Therefore it follows from what was proved in Section **1.1** of this chapter that

$$f(z) = \int_{-\infty}^{\infty} \frac{d\sigma(u)}{u-z},$$

where $\sigma(u)$ is a non-decreasing function of bounded variation, and we conclude from [3.8a] that

$$s_0 = \int_{-\infty}^{\infty} d\sigma(u).$$

Further it follows from [3.8] that

$$\lim_{y \to \infty} (iy)^{2m+1} \left\{ f(iy) + \frac{s_0}{iy} + \frac{s_1}{(iy)^2} + \ldots + \frac{s_{2m-1}}{(iy)^{2m}} \right\} = -s_{2m}$$

for $m = 1, 2, \ldots, n$. Therefore assuming that the equations

$$s_k = \int_{-\infty}^{\infty} u^k \, d\sigma(u) \quad (k = 0, 1, 2, \ldots, 2m-2; \; m \leqq n)$$

have been proved we can write down the relation

$$\lim_{y \to \infty} \left\{ \int_{-\infty}^{\infty} \frac{(iy)^2 u^{2m-1}}{u-iy} \, d\sigma(u) + iys_{2m-1} \right\} = -s_{2m},$$

from which it follows in particular that

$$\lim_{y \to \infty} \int_{-\infty}^{\infty} \frac{iyu^{2m-1}}{u-iy} \, d\sigma(u) = -s_{2m-1}.$$

Splitting off the real parts we obtain the equations

$$s_{2m-1} = \lim_{y \to \infty} \int_{-\infty}^{\infty} \frac{y^2 u^{2m-1}}{u^2+y^2} \, d\sigma(u), \qquad \ldots [3.9]$$

$$s_{2m} = \lim_{y \to \infty} \int_{-\infty}^{\infty} \frac{y^2 u^{2m}}{u^2+y^2} \, d\sigma(u). \qquad \ldots [3.10]$$

We find from [3.10] that

$$s_{2m} = \int_{-\infty}^{\infty} u^{2m} \, d\sigma(u).$$

Therefore the integral

$$\int_{-\infty}^{\infty} |u|^{2m-1} \, d\sigma(u)$$

exists and then equation [3.9] shows that

$$s_{2m-1} = \int_{-\infty}^{\infty} u^{2m-1} \, d\sigma(u).$$

Thus induction may be used and the second statement of the theorem is also proved.

If the relation [3.7] is true for any n ($= 0, 1, 2, \ldots$), the series

$$-\frac{s_0}{z} - \frac{s_1}{z^2} - \frac{s_2}{z^3} - \ldots - \frac{s_{2n}}{z^{2n+1}} - \ldots$$

is an asymptotic expansion of $f(z)$.

It follows from this that the moment problem is equivalent to the problem of finding a function $f(z) \in N$ which has a given asymptotic expansion. This latter can be interpreted† as a certain expression for the interpolation of $f(z_\alpha) = w_\alpha$.

2.2. Using this result Nevanlinna's theorem can be proved, which gives a description of all solutions of the moment problem in the indeterminate case.

† This will be discussed in more detail below, see Section 3.6.

THEOREM **3.2.2.** (NEVANLINNA[1]). *Assume that*

$$\left\| \begin{array}{cc} A(z) & C(z) \\ B(z) & D(z) \end{array} \right\|$$

is a Nevanlinna matrix corresponding to an indeterminate moment problem; in this case the formula

$$\int_{-\infty}^{\infty} \frac{d\sigma(u)}{u-z} = -\frac{A(z)\phi(z)-C(z)}{B(z)\phi(z)-D(z)} \qquad \ldots[3.11]$$

establishes a one-to-one correspondence between V, *the aggregate of all solutions* $\sigma(u)$ *of the moment problem in question and the aggregate of all functions* $\phi(z)$ *of the class* N *augmented by the constant* ∞ *(finite real constants are included in the class* N *by definition).*

(To determine the function $\sigma(u)$ from the function $\phi(z)$ one can use the Stieltjes-Perron inversion formula†.)

Proof. We consider a mapping given by the formula

$$w = -\frac{A\ z)t-B(z)}{B(z)t-D(z)}.$$

For fixed real z this formula transforms the half-plane Im $t > 0$ into the half plane Im $w > 0$. Therefore if the point t moves along the real axis from $-\infty$ to $+\infty$ the point w also moves along the real axis from $-\infty$ to $+\infty$. Therefore if Im $z > 0$, when the real axis of the t-plane goes over into the circle $K_\infty(z)$, which lies entirely within the half-plane Im $w > 0$, the movement of the point t along the real axis from $-\infty$ to $+\infty$ will correspond to the movement of w over the circle $K_\infty(z)$ in the counter-clockwise sense. Therefore if Im $z > 0$ the half-plane Im $t \geq 0$ will go over into the interior of the circle $K_\infty(z)$ (and not into its exterior). Similarly for Im $z < 0$ the half-plane Im $t \leq 0$ will go over into the interior of the circle $K_\infty(z)$. This mapping is a one-to-one correspondence. Therefore if Im $z \neq 0$ and $w \in K_\infty(z)$ the inequality

$$\frac{\text{Im } t}{\text{Im } z} \geq 0$$

certainly holds.

We shall start from a certain solution $\sigma(u)$ of our moment problem and take the integral

$$\int_{-\infty}^{\infty} \frac{d\sigma(u)}{u-z}$$

_____
† See Addenda and Problems to Chapter 3, example 1.

as the quantity $w \in K_\infty(z)$. Then equation [3.11] defines $\phi(z)$ as a certain new function of z in each of the half planes Im $z > 0$ and Im $z < 0$. If this function is not an infinite constant it is meromorphic in each of the two half-planes and from what has been said above we have

$$\frac{\text{Im } \phi(z)}{\text{Im } z} \geqq 0.$$

Therefore in the case when $\phi(z)$ is not an infinite constant it is regular in each of the half-planes Im $z > 0$ and Im $z < 0$ and therefore belongs to the class N.†

Thus one part of the theorem is proved.

We proceed to prove the second half of the theorem. We are given a function $\phi(z) \in N$ (the case when $\phi(z) = $ const need not be discussed) and we must prove that the expression

$$w = -\frac{A(z)\phi(z) - C(z)}{B(z)\phi(z) - D(z)}$$

may be represented in the form

$$w = \int_{-\infty}^{\infty} \frac{d\sigma(u)}{u - z},$$

where $\sigma(u)$ is some solution of our moment problem. We put

$$w_n = -\frac{A_n(z)\phi(z) - C_n(z)}{B_n(z)\phi(z) - D_n(z)},$$

so that w is the limit of w_n for $n \to \infty$ and in fact is a limit which is attained uniformly in any finite closed region which has no points in common with the real axis.

The function w_n belongs to the class N and the rational fractions $A_n(z)/B_n(z)$, and $C_n(z)/D_n(z)$ permit the expansion

$$\frac{A_n(z)}{B_n(z)} = \frac{s_0}{z} + \frac{s_1}{z^2} + \ldots + \frac{s_{2n-2}}{z^{2n-1}} + O\left(\frac{1}{z^{2n}}\right),$$

$$\frac{C_n(z)}{D_n(z)} = \frac{s_0}{z} + \frac{s_1}{z^2} + \ldots + \frac{s_{2n-2}}{z^{2n-1}} + O\left(\frac{1}{z^{2n}}\right).$$
...[3.12]

On the other hand w_n can be represented in the form

$$w_n = w_n(z) = -\frac{A_n(z)}{B_n(z)} - \frac{1}{B_n(z)[B_n(z)\phi(z) - D_n(z)]}. \quad \ldots[3.13]$$

† The equality $\overline{\phi(z)} = \phi(\bar{z})$ is a consequence of the fact that the entire functions $A(z)$, $B(z)$, $C(z)$, $D(z)$ are real.

It is easy to utilize this expression if the polynomial $B_n(z)$ is of exactly n-th degree and if the function $\phi(z)$ satisfies the condition.

$$\lim_{y \to \infty} \frac{\phi(iy)}{iy} > 0, \qquad \dots[3.14]$$

which means that the constant μ in the integral representation of $\phi(z)$ is positive. Indeed, in this case it follows immediately from [3.12] and [3.13] that for $y \to \infty$ we have

$$w_n(iy) = -\frac{s_0}{iy} - \frac{s_1}{(iy)^2} - \dots - \frac{s_{2n-2}}{(iy)^{2n-1}} + O\left(\frac{1}{y^{2n}}\right). \quad \dots[3.12a]$$

If the degree of the polynomial $B_n(z)$ is less than n, then the degree of the polynomial $B_{n+1}(z)$ is certainly $n+1$ and therefore the relation [3.12a] is true for an infinitely increasing sequence of suffixes n. Since from Theorem 3.2.1 we have

$$w_n(z) = \int_{-\infty}^{\infty} \frac{d\sigma_n(u)}{u-z},$$

where $\sigma_n(u)$ is a certain solution of the truncated problem, we can go to the limit and using Helly's theorem find that

$$w = \int_{-\infty}^{\infty} \frac{d\sigma(u)}{u-z},$$

where $\sigma(u)$ is a solution of the full moment problem.

It remains to discuss the case when condition [3.14] is not satisfied.

In this case we introduce the function $\phi_\alpha(z) = \alpha z + \phi(z)$ where α is a positive constant. The function $\phi_\alpha(z)$ now does satisfy condition [3.14] and it certainly belongs to the class N. Therefore from what has been proved already we have

$$-\frac{A(z)\phi_\alpha(z) - C(z)}{B(z)\phi_\alpha(z) - D(z)} = \int_{-\infty}^{\infty} \frac{d\sigma^{(\alpha)}(u)}{u-z},$$

where the function $\sigma^{(\alpha)}(u)$ is a solution of the full moment problem for any $\alpha > 0$. To complete the proof it is sufficient to go to the limit ($\alpha \to 0$) again and to use Helly's theorem.

Theorem 3.2.2 leads to an important conclusion concerning N-extremal solutions in the indeterminate case. It follows from Section 4.2 of Chapter 2 that if one puts a real constant into equation [3.11] in place of the function $\phi(z)$ (the constant may be finite or infinite) the corresponding function $\sigma(u)$ will be an N-extremal solution of the moment problem. We can now assert, by Theorem 3.2.2, that all

N-extremal solutions can be obtained in this way. Further, it follows from formula [3.11], that in the indeterminate case the points of increase of an N-extremal solution are the zeros of an entire function of minimal exponential type (namely the function $B(z)t - D(z)$, if $t \neq \infty$, and of the function $B(z)$ if $t = \infty$).

3. An algorithm for consecutive linear fractional transformations

3.1. The case of the general interpolational problem, to which the power moment problem leads, is a limiting one. In the previous section we made use of the apparatus of integral representations in this limiting case. This apparatus will also be used in the basic case of the interpolational problem and in the third sub-section of the present section we shall develop one of the variants of its application to the Nevanlinna-Pick problem in the class N. However, there exists also another method for treating these problems. In it one uses the algorithm of consecutive linear fractional transformations, based on well-known theorems in the theory of functions (on the so-called Schwartz and Julia-Carathéodory Lemmas). This algorithm was used in a masterly fashion by NEVANLINNA[1, 2] and was also used somewhat earlier in a different variant by SCHUR[2], in connection with a related problem. In the present section we indicate the nature of this algorithm.

3.2. SCHUR'S COEFFICIENT PROBLEM: *Find necessary and sufficient conditions for the function*

$$\phi(z) = \alpha_0 + \alpha_1 z + \alpha_2 z^2 + \ \ldots \qquad \ldots [3.15]$$

to belong to the class S.

In the first place we note that the inequality

$$|\alpha_0| \leqq 1 \qquad \ldots [3.16]$$

must hold since

$$\alpha_0 = \phi(0).$$

By the principle of maximum modulus the equality sign in equation [3.16] is possible only in the case when

$$\phi(z) \equiv \alpha_0.$$

Assuming that $|\alpha_0| < 1$, we construct the function

$$\phi_1(z) = \frac{1}{z} \frac{\phi(z) - \alpha_0}{1 - \bar{\alpha}_0 \phi(z)},$$

so that

$$\phi(z) = \frac{\alpha_0 + z\phi_1(z)}{1 + \bar{\alpha}_0 z\phi_1(z)}. \qquad \ldots[3.17]$$

From the equality

$$\phi(0) = \alpha_0 \quad (|\,\alpha_0\,| < 1)$$

and from Schwartz' lemma it follows that $\phi_1(z)$ is regular in the circle $|z| < 1$ and satisfies the equality $|\,\phi_1(z)\,| \leqq 1$. Conversely for any function $\phi_1(z)$ satisfying these conditions the function $\phi(z)$ defined by equation [3.17] will be regular in the circle $|z| < 1$ and will satisfy the relations

$$|\,\phi(z)\,| < 1$$

and

$$\phi(0) = \alpha_0.$$

Thus equation [3.17] gives the general form of the function $\phi(z)$ satisfying these two conditions. The expansion of $\phi_1(z)$ must have the form

$$\phi_1(z) = \alpha_0^{(1)} + \alpha_1^{(1)}z + \alpha_2^{(1)}z^2 + \ldots,$$

where the coefficients $\alpha_k^{(1)}$ are expressible in a definite form in terms of the initial coefficients α_k. For instance

$$\alpha_0^{(1)} = \frac{\alpha_1}{1 - |\,\alpha_0\,|^2}.$$

Now we are again led to the inequality

$$|\,\alpha_0^{(1)}\,| \leqq 1.$$

If here we have the $=$ sign, i.e. if

$$|\,\alpha_1\,| = 1 - |\,\alpha_0\,|^2,$$

the function $\phi_1(z)$ is unique, namely

$$\phi_1(z) \equiv \alpha_0^{(1)},$$

and with it the function $\phi(z)$ is unique:

$$\phi(z) = \frac{\alpha_0 + \alpha_0^{(1)}z}{1 + \bar{\alpha}_0 \alpha_0^{(1)}z}.$$

Assuming that $|\,\alpha_0^{(1)}\,| < 1$ we introduce the function

$$\phi_2(z) = \frac{1}{z}\frac{\phi_1(z) - \alpha_0^{(1)}}{1 - \bar{\alpha}_0^{(1)}\phi_1(z)},$$

such that

$$\phi_1(z) = \frac{\alpha_0^{(1)} + z\phi_2(z)}{1 + \overline{\alpha}_0^{(1)} z\phi_2(z)}.$$

If we insert this expression into equation [3.17] we obtain the general form of a function $\phi(z)$ which is regular in the region $|z| < 1$ and satisfies the relations

$$|\phi(z)| < 1, \quad \phi(0) = \alpha_0, \quad \frac{\phi'(0)}{1!} = \alpha_1,$$

namely

$$\phi(z) = \frac{\alpha_0 + z \dfrac{\alpha_0^{(1)} + z\phi_2(z)}{1 + \overline{\alpha}_0^{(1)} z\phi_2(z)}}{1 + \overline{\alpha}_0 z \dfrac{\alpha_0^{(1)} + z\phi_2(z)}{1 + \overline{\alpha}_0^{(1)} z\phi_2(z)}},$$

where $\phi_2(z)$ is an arbitrary function that is regular in the circle $|z| < 1$ and which satisfies the inequality $|\phi_2(z)| \leqq 1$.

The further continuation of the procedure is clear. However, in order to obtain perspicuous results one has to perform certain calculations, namely to find the values at $z = 0$ of the consecutive functions $\phi_n(z)$:

$$\phi_n(0) = \alpha_0^{(n)} = a_n \quad (a_0 = \alpha_0; \; n = 0, 1, 2, \ldots).$$

With the aid of these numbers a_n (they are called *Schur's parameters*) the answer to the question posed is simply stated:

THEOREM **3.2.2.** *In order that the series [3.15] should represent a function of class* S *it is necessary and sufficient that one of the following two statements be true: either*

$$|a_k| < 1 \quad (k = 0, 1, 2, \ldots),$$

or

$$|a_k| < 1 \quad (k = 0, 1, 2, \ldots, n-1), \quad |a_n| = 1$$

and $\phi(z)$ *is a rational fraction of degree* n.

We shall not dwell here on the evaluation of the Schur parameters and will only note that the algorithm described leads to an expansion of $\phi_n(z)$ into certain continued fractions. If

$$0 < |a_k| < 1 \quad (k = 0, 1, 2, \ldots),$$

these continued fractions have the form

$$\phi_n(z) = \cfrac{a_n}{1 - \cfrac{\dfrac{a_{n+1}}{a_n}(1 - |a_n|^2)z}{1 + \dfrac{a_{n+1}}{a_n}z - \cfrac{\dfrac{a_{n+2}}{a_{n+1}}(1 - |a_{n+1}|^2)z}{1 + \dfrac{a_{n+2}}{a_{n+1}}z - \cdots}}}$$

and converge uniformly for $|z| \leqq r < 1$.

The detailed investigation of many facts related to this is due to GERONIMUS[2, 4].

3.3. THE NEVANLINNA-PICK PROBLEM IN THE CLASS N: *given a set of numbers* $\{w_\alpha\}$, *find necessary and sufficient conditions for a function* $w = f(z) \in N$, *to exist, which satisfies the equations*

$$f(z_\alpha) = w_\alpha \quad (z_\alpha \in Z),$$

where $Z = \{z_\alpha\}$ *is a given set of points in the half plane* Im $z > 0$.†

Assuming that the desired function exists we write its integral representation

$$f(z) = \mu z + \nu + \int_{-\infty}^{\infty} \frac{1 + uz}{u - z} \, d\sigma(u) \quad (\mu \geqq 0).$$

By virtue of this representation we have for any two points $z_\alpha, z_\beta \in Z$:

$$\frac{w_\alpha - \bar{w}_\beta}{z_\alpha - \bar{z}_\beta} = \mu + \int_{-\infty}^{\infty} \frac{1 + u^2}{(u - z_\alpha)(u - \bar{z}_\beta)} \, d\sigma(u).$$

Therefore, taking an aribtrary selection of points $z_{\alpha_0}, z_{\alpha_1}, \ldots z_{\alpha_n} \in Z$ and forming the expression

$$\sum_{j, k=0}^{n} \frac{w_{\alpha_j} - \bar{w}_{\alpha_k}}{z_{\alpha_j} - \bar{z}_{\alpha_k}} \, \zeta_j \bar{\zeta}_k, \qquad \ldots [3.18]$$

we find that it is equal to

$$\mu \left| \sum_0^n \zeta_k \right|^2 + \int_{-\infty}^{\infty} \left| \sum_0^n \frac{\zeta_k}{u - z_{\alpha_k}} \right| (1 + u^2) \, d\sigma(u) \geqq 0.$$

---

† PICK[1, 2] gave the solution of this problem for the case when the set Z is finite, NEVANLINNA[2] extended this to the case of a denumerable set and discussed the problem of determinateness. For arbitrary sets Z the problem was solved in a paper by KREIN and REKHTMAN[1].

Thus the non-negative property of all possible forms [*3.18*] is a necessary condition for the existence of the required functions. In other words, the first statement of the following theorem has been proved.

THEOREM 3.3.3. *For the existence of a function* $f(z) \in N$, *which satisfies the condition*

$$f(z_\alpha) = w_\alpha \quad (z_\alpha \in Z),$$

where Z *is a given point set in the half plane* Im $z > 0$ *it is necessary and sufficient that all forms*

$$\sum_{j,\,k=0}^{n} \frac{w_{\alpha_j} - \bar{w}_{\alpha_k}}{z_{\alpha_j} - \bar{z}_{\alpha_k}} \xi_j \bar{\xi}_k$$

should be non-negative. If any one of these forms is singular, then the function $f(z)$ *is unique and equal to a real rational fraction.*

We now turn to the sufficiency proof. We assume that without loss of generality one of the points z_α (say $z_{\alpha_0} = z_0$) is i, and also that the corresponding value of $f(z)$ (i.e. w_0) is a purely imaginary number. One can always reduce things to this case by integral linear transformations

$$z = x_0 + y_0 \zeta, \quad w = u_0 + \omega,$$

where $z_0 = x_0 + iy_0$, and $u_0 = \operatorname{Re} w_0$.

We introduce G, the aggregate of functions of the variable u $(-\infty < u < \infty)$ of the form

$$\phi_\alpha(u) = \frac{(1+u^2)(u-x_\alpha)}{(u-x_\alpha)^2 + y_\alpha^2} - u,$$

$$\psi_\alpha(u) = \frac{(1+u^2)y_\alpha}{(u-x_\alpha)^2 + y_\alpha^2},$$

where $z_\alpha = x_\alpha + iy_\alpha$ ranges over the set Z. By virtue of our condition $\phi_0(u) \equiv 0$, $\psi_0(u) \equiv 1$.

We form the linear hull $\mathfrak{M}$ of the set G and determine the linear functional $\mathfrak{P}$ on $\mathfrak{M}$ with the aid of the identifications

$$\mathfrak{P}\{\phi_\alpha(u)\} = \frac{w_\alpha + \bar{w}_\alpha}{2}, \quad \mathfrak{P}\{\psi_\alpha(u)\} = \frac{w_\alpha - \bar{w}_\alpha}{2i}.$$

Since

$$\phi_\alpha(u) + i\psi_\alpha(u) = \frac{1 + uz_\alpha}{u - z_\alpha},$$

we have

$$\mathcal{P}\left\{\frac{1+uz_\alpha}{u-z_\alpha}\right\} = w_\alpha, \quad \mathcal{P}\left\{\frac{1-u\bar{z}_\alpha}{u-\bar{z}_\alpha}\right\} = \bar{w}_\alpha.$$

It is not difficult to verify that the functional $\mathcal{P}$ is non-negative. Indeed, if a certain function $\Phi(u) \in \mathfrak{M}$ satisfies the inequality

$$\Phi(u) \geqq 0 \quad (-\infty < u < \infty),$$

it can be represented in the form

$$\Phi(u) = \sum_0^n \left[A_k \frac{1+uz_{\alpha_k}}{u-z_{\alpha_k}} + \bar{A}_k \frac{1+u\bar{z}_{\alpha_k}}{u-\bar{z}_{\alpha_k}} \right]$$

$$= \frac{R(u)}{|\,(u-z_{\alpha_1})(u-z_{\alpha_2})\ldots(u-z_{\alpha_n})\,|^2},$$

where $R(u)$ is a polynomial of degree $\leqq 2n$ satisfying the inequality

$$R(u) \geqq 0 \quad (-\infty < u < \infty).$$

Therefore

$$R(u) = \left| \sum_0^n \eta_k u^k \right|^2$$

and thus

$$\Phi(u) = \left| \frac{\sum_0^n \eta_k u^k}{(u-z_{\alpha_1})(u-z_{\alpha_2})\ldots(u-z_{\alpha_n})} \right|^2.$$

But it is easy to see that the rational fraction

$$\frac{\sum_0^n \eta_k u^k}{(u-z_{\alpha_1})(u-z_{\alpha_2})\ldots(u-z_{\alpha_n})}$$

can be represented, uniquely, in the form

$$\sum_{k=0}^n \xi_k \frac{u-i}{u-z_{\alpha_k}}.$$

Therefore

$$\Phi(u) = \left| \sum_{k=0}^n \xi_k \frac{u-i}{u-z_{\alpha_k}} \right|^2 = \sum_{j,\,k=0}^n \frac{1+u^2}{(u-z_{\alpha_j})(u-\bar{z}_{\alpha_k})} \xi_k \bar{\xi}_k$$

$$= \sum_{j,\,k=0}^n \frac{1}{z_\alpha - \bar{z}_{\alpha_k}} \left[\frac{1+uz_{\alpha_j}}{u-z_\alpha} - \frac{1+u\bar{z}_{\alpha_k}}{u-\bar{z}_{\alpha_k}} \right] \xi_j \bar{\xi}_k$$

and so

$$\mathcal{P}\{\Phi(u)\} = \sum_{j,\,k=0}^{n} \frac{w_{\alpha_j} - \overline{w}_{\alpha_k}}{z_{\alpha_j} - \overline{z}_{\alpha_k}} \, \xi_j \overline{\xi}_k \geqq 0.$$

Thus we have proved that the functional $\mathcal{P}$ is non-negative. We now note that for any of the functions $\phi_\alpha(u)$, $\psi_\alpha(u)$ we have

$$\frac{\phi_\alpha(u)}{\psi_0(u)} = \phi_\alpha(u) \to x_\alpha, \qquad \frac{\psi_\alpha(u)}{\psi_0(u)} = \psi_\alpha(u) \to y_\alpha \quad (\pm u \to \infty).$$

Therefore the conditions of Theorem **2.6.6** are satisfied and so there exists a constant $\mu \geqq 0$ and a non-decreasing function of bounded variation $\sigma(u)\,(-\infty \leqq u \leqq \infty)$ such that

$$\mathcal{P}\{\phi_\alpha(u)\} = \mu x_\alpha + \int_{-\infty}^{\infty} \phi_\alpha(u)\, d\sigma(u),$$

$$\mathcal{P}\{\psi_\alpha(u)\} = \mu y_\alpha + \int_{-\infty}^{\infty} \psi_\alpha(u)\, d\sigma(u).$$

But this means that

$$w_\alpha = \mathcal{P}\{\phi_\alpha(u) + i\psi_\alpha(u)\} = \mu z_\alpha + \int_{-\infty}^{\infty} \frac{1 + u z_\alpha}{u - z_\alpha}\, d\sigma(u),$$

or in other words that a function

$$f(z) = \mu z + \int_{-\infty}^{\infty} \frac{1 + uz}{u - z}\, d\sigma(u) \in \mathbf{N}$$

has been found for which

$$f(z_\alpha) = w_\alpha \quad (z_\alpha \in \mathbf{Z}).$$

Going over to the proof of the last statement we denote by $z_1, z_2, \ldots, z_n$ that selection of points from the set $\mathbf{Z}$ which generates the lowest order singular form, and assume that

$$\sum_{j,\,k=1}^{n} \frac{w_j - \overline{w}_k}{z_j - \overline{z}_k} \, \gamma_j \overline{\gamma}_k = 0,$$

where all the γ_j are different from zero. Assume further that

$$f(z) = \mu z + \int_{-\infty}^{\infty} \frac{1 + uz}{u - z}\, d\sigma(u)$$

is some function satisfying the condition

$$f(z_\alpha) = w_\alpha \quad (z_\alpha \in \mathbf{Z}).$$

In this case we have

$$\sum_{j,\,k=1}^{n} \frac{w_j - \bar{w}_k}{z_j + \bar{z}_k} \gamma_j \bar{\gamma}_k = \mu \left| \sum_1^n \gamma_j \right|^2 + \int_{-\infty}^{\infty} \left| \sum_1^n \frac{\gamma_j}{u - z_j} \right|^2 (1 + u^2) \, d\sigma(u)$$

$$= \mu \left| \sum_1^n \gamma_j \right|^2 + \int_{-\infty}^{\infty} \frac{|R(u)|^2}{|(u - z_1) \dots (u - z_n)|^2} (1 + u^2) \, d\sigma(u),$$

where $R(u)$ is some polynomial with the leading term $\sum_1^n \gamma_j u^{n-1}$. Since the left hand side of this equation is zero both terms of the right hand side are zero and therefore only the zeros of the polynomial $R(u)$ can be points of increase of $\sigma(u)$, and the number of these is $\leqq n - 1$. Therefore $f(z)$ has the form

$$f(z) = \mu z + \sum_1^{n-1} \mu_k \frac{1 + u_k z}{u_k - z}$$

and the following two cases are possible:

1) $\sum_1^n \gamma_j \neq 0, \quad \mu = 0, \quad \mu_k > 0, \quad k = 1, 2, \dots, n-1;$

2) $\sum_1^n \gamma_j = 0, \quad \mu > 0, \quad \mu_k > 0, \quad k = 1, 2, \dots, n-2, \quad \mu_{n-1} = 0.$

It is not difficult to verify that in both cases the coefficients μ, and μ_k are uniquely determined.†

Thus the last statement of the theorem is also proved.

3.4. If we had been discussing the Nevanlinna-Pick problem in the class C we would have obtained for the pair of points $z_\alpha, z_\beta \in Z$, where now Z lies within the circle $|z| < 1$, by virtue of the Riesz-Herglotz equation that

$$\frac{w_\alpha + \bar{w}_\beta}{1 - z_\alpha \bar{z}_\beta} = \frac{1}{2\pi(1 - z_\alpha \bar{z}_\beta)} \int_{-\pi}^{\pi} \left\{ \frac{e^{i\theta} + z_\alpha}{e^{i\theta} - z_\alpha} + \frac{e^{-i\theta} + \bar{z}_\beta}{e^{-i\theta} - \bar{z}_\beta} \right\} d\sigma(\theta)$$

$$= \frac{1}{\pi} \int_{-\pi}^{\pi} \frac{1}{(e^{i\theta} - z_\alpha)(e^{-i\theta} - \bar{z}_\beta)} d\sigma(\theta)$$

and therefore we would have concluded that the form

$$\sum_{j,\,k=1}^{n} \frac{w_{\alpha_j} + \bar{w}_{\alpha_k}}{1 - z_{\alpha_j} \bar{z}_{\alpha_k}} \xi_j \bar{\xi}_k \qquad \qquad \dots [3.18a]$$

† Indeed these $n-1$ coefficients satisfy the system of equations

$$\mu + \sum_{k=1}^{n-1} \frac{1}{u_k - z_j} \frac{1 + u_k^2}{u_k - z_1} \mu_k = \frac{w_j - w_1}{z_j - z_1} \quad (j = 2, 3, \dots, n),$$

which has a non-vanishing determinant (see e.g. POLYA and SZEGO[1]).

is non-negative;† this is the necessary condition. That the condition is also sufficient can be proved in the same way as the corresponding result for the class N. However it is not necessary to carry through this proof because the result for class C can be obtained by means of simple transformation from the corresponding result for class N.

3.5. We shall dwell a little longer on the Nevanlinna-Pick problem for class N in the case when the set Z is denumerable and therefore the conditions consist in requiring that

$$f(z_k) = w_k \quad (k = 1, 2, 3, \ldots). \qquad \ldots[3.19]$$

We shall in particular make clear the form taken in this case by the algorithm of successive fractional linear transformations.

We take the first of conditions [3.19]. In order that a function $f(z) \in N$ should exist which satisfies this condition $(f(z_1) = w_1)$, it is necessary that the inequality Im $w_1 \geqq 0$ should be satisfied. If the = sign holds here, the required function can only be a constant, w_1. If on the other hand Im $w_1 > 0$ we put

$$\frac{f(z) - w_1}{f(z) - \bar{w}_1} : \frac{z - z_1}{z - \bar{z}_1} = \phi_1(z). \qquad \ldots[3.20]$$

The function $\phi_1(z)$ so obtained is regular in the half plane Im $z > 0$ and by virtue of Schwartz' lemma it satisfies the inequality $| \phi_1(z) | \leqq 1$. Conversely, with any choice of $\phi_1(z)$ possessing these properties equation [3.20] defines a function $f(z)$ belonging to the class N and satisfying the condition $f(z_1) = w_1$.

We now take an arbitrary constant γ (Im $\gamma > 0$) and put

$$\phi_1(z) = \frac{f_1(z) - \gamma}{f_1(z) - \bar{\gamma}}. \qquad \ldots[3.21]$$

Then $f_1(z)$ will belong to the class N and conversely for any arbitrary choice of a function $f_1(z) \in N$ equation [3.21] gives a function $\phi_1(z)$ which is regular in the region Im $z > 0$ and satisfies the inequality $| \phi_1(z) | \leqq 1$.

† It is worth while noting how the structure of the forms reflects the class (N or C) under discussion. Take the coefficient of the diagonal term. In the one case it is

$$\frac{w_\alpha - \bar{w}_\alpha}{z_\alpha - \bar{z}_\alpha} = \frac{\text{Im } w_\alpha}{\text{Im } z_\alpha},$$

in the other

$$\frac{w_\alpha + \bar{w}_\alpha}{1 - |z_\alpha|^2} = \frac{2\text{Re } w_\alpha}{1 - |z_\alpha|^2}.$$

We find from the equations [3.20] and [3.21] that

$$f(z) = \frac{f_1(z)[\operatorname{Im}(z_1\bar{w}_1) - z\operatorname{Im}\bar{w}_1] + z\operatorname{Im}(\gamma\bar{w}_1) - \operatorname{Im}(z_1\gamma\bar{w}_1)}{f_1(z)\operatorname{Im} z_1 + z\operatorname{Im}\gamma - \operatorname{Im}(\gamma z_1)}$$

and

$$f_1(z) = \frac{f(z)[\operatorname{Im}(\gamma z_1) - z\operatorname{Im}\gamma] + z\operatorname{Im}(\gamma\bar{w}_1) - \operatorname{Im}(z_1\gamma\bar{w}_1)}{f(z)\operatorname{Im} z_1 - \operatorname{Im}(z_1\bar{w}_1) + z\operatorname{Im}\bar{w}_1}.$$

The first of these formulae gives the general form of all functions $f(z) \in N$ which satisfy only the first of the conditions [3.19]. The second allows one to formulate the remaining conditions in terms of the function $f_1(z)$:

$$f_1(z_k) = w_k^{(1)} \quad (k = 2, 3, \ldots).$$

One then performs the next step. After n operations one obtains for $f(z)$ the representation

$$f(z) = \frac{-\alpha_n(z) + \beta_n(z)f_n(z)}{\gamma_n(z) - \delta_n(z)f_n(z)},$$

where $\alpha_n(z)$, $\beta_n(z)$, $\gamma_n(z)$, $\delta_n(z)$ are some polynomials of degree n, the calculation of which we shall not discuss.

3.6. If in the Nevanlinna-Pick problem for the class N all the points z_k were to coalesce into the single point z_0, lying within the region $\operatorname{Im} z > 0$, condition [3.19] would be replaced by the requirement that in the neighbourhood of this point z_0 one should have the expansion

$$f(z) = \sum_0^\infty C_k(z - z_0)^k,$$

where the C_k are given numbers. If the point z_0 were moved to infinity then one would have in place of the Taylor series of the last equation the asymptotic expansion

$$f(z) \sim -\frac{s_0}{z} - \frac{s_1}{z^2} - \frac{s^2}{z^3} - \cdots,$$

which must hold for $z \to \infty$ in the range of angles† $\delta \leq \arg z \leq \pi - \delta$ for any positive $\delta < \frac{1}{2}\pi$.

The problem of finding functions $f(z) \in N$ with a given asymptotic expansion of the form indicated is equivalent to the moment problem. Therefore it is natural to examine what form is taken on by the algorithm of successive linear fractional transformations in this case. The

† In the following all other asymptotic expansions will be assumed to be applicable in the same range of angles as also will the symbol $o(z^m)$.

basis of the algorithm in this case is not Schwartz' lemma but the following one:

LEMMA **3.3.6.** *If* $\Phi(z) \in N$ *and*

$$\Phi(z) \sim -\frac{A}{z} - \frac{B}{z^2},$$

where A and B are real numbers, then either $A = 0$ *and* $\Phi(z) \equiv 0$ *or* $A > 0$ *and*

$$\Phi(z) = -\frac{A}{z - \dfrac{B}{A} + \Phi^*(z)},$$

where $\Phi^*(z) \in N$ *and* $\Phi^*(z) = o(1)$.

This lemma can be proved with the aid of the Julia-Carathéodory lemma. We give a proof of it based on the Nevanlinna formula.

By virtue of the conditions of the lemma and of Theorem **3.2.1** the function $\Phi(z)$ can be represented in the form

$$\Phi(z) = \int_{-\infty}^{\infty} \frac{d\omega(u)}{u - z},$$

where $\omega(u)$ is a non-decreasing function and

$$A = \int_{-\infty}^{\infty} d\omega(u).$$

Hence it follows that $A \geqq 0$ and also that for $A = 0$

$$\Phi(z) \equiv 0.$$

Assuming that $A > 0$, we put

$$\Phi(z) = -\frac{A}{z - \dfrac{B}{A} + \Phi^*(z)}. \qquad \ldots[3.22]$$

In this case

$$\Phi^*(z) = -z + \frac{B}{A} + \frac{A}{\displaystyle\int_{-\infty}^{\infty} \frac{d\omega(u)}{z - u}}.$$

Hence one sees that $\Phi^*(z)$ is regular for $y = \operatorname{Im} z \neq 0$. Further, we have

$$\operatorname{Im} \Phi^*(z) = -y + y \int_{-\infty}^{\infty} \frac{d\omega(u)}{|z-u|^2} \frac{A}{\left| \int_{-\infty}^{\infty} \frac{d\omega(t)}{z-t} \right|^2}.$$

But by virtue of the Schwartz-Bunyakovskii inequality we have

$$\left| \int_{-\infty}^{\infty} \frac{d\omega(t)}{z-t} \right|^2 \leq \int_{-\infty}^{\infty} d\omega(t) \int_{-\infty}^{\infty} \frac{d\omega(u)}{|z-u|^2}.$$

Therefore

$$\frac{\operatorname{Im} \Phi^*(z)}{\operatorname{Im} z} \geq 0 \quad (\operatorname{Im} z \neq 0).$$

Finally it follows from the condition of the lemma and from [3.22] that

$$\Phi^*(z) = -z + \frac{B}{A} + \frac{A}{\dfrac{A}{z} + \dfrac{B}{z^2} + o\left(\dfrac{1}{z^2}\right)} = o(1).$$

Thus the lemma is proved.

We now turn to the function

$$f(z) \sim -\frac{s_0}{z} - \frac{s_1}{z^2} - \frac{s_2}{z^3} - \dots .$$

From Lemma **3.3.6** we conclude that the inequality $s_0 \geq 0$ must hold. If the equality sign holds, we have $f(z) \equiv 0$. Assuming that $s_0 > 0$ we put

$$f(z) = -\frac{s_0}{z - a_0 + f_1(z)} \qquad \left(a_0 = \frac{s_1}{s_0} \right); \qquad \dots [3.23]$$

then $f_1(z) \in N$ and $f_1(z) = o(1)$. Using the asymptotic expansion for $f(z)$ we can construct an asymptotic expansion for $f_1(z)$:

$$f_1(z) \sim -\frac{s_0^{(1)}}{z} - \frac{s_1^{(1)}}{z^2} - \frac{s_2^{(1)}}{z^3} - \dots .$$

Here the inequality $s_0^{(1)} \geq 0$ must hold, which is a condition imposed on the moments s_0, s_1 and s_2. The = sign in this inequality is possible only for $f_1(z) \equiv 0$. Disregarding this case we put $s_0^{(1)} = b_0^2$ and, again applying the lemma, we find that

$$f_1(z) = -\frac{b_0^2}{z - a_1 + f_2(z)} \qquad \left(a_1 = \frac{s_1^{(1)}}{s_0^{(1)}} \right), \qquad \dots [3.24]$$

where $f_2(z) \in N$ and $f_2(z) = o(1)$.   Here one finds for $f_2(z)$ a certain asymptotic expansion

$$f_2(z) \sim -\frac{s_0^{(2)}}{z} - \frac{s_1^{(2)}}{z^2} - \frac{s_2^{(2)}}{z^3} - \cdots .$$

The further continuation of the procedure is evident.   Formula [3.23] with an arbitrary function $f_1(z) \in N$ satisfying the condition $f_1(z) = o(1)$ gives a general form of a function $f(z) \in N$ with the required first two terms in its asymptotic expansion.   To obtain this formula is the first step in the algorithm.   Inserting [3.24] into [3.23] we obtain the general form of a function $f(z) \in N$ with the required first four terms in the asymptotic expansion, namely

$$f(z) = -\cfrac{s_0}{z - a_0 - \cfrac{b_0^2}{z - a_1 + f_2(z)}}.$$

The construction of this formula is the second step.   After n such steps we get to the general form of all functions $f(z) \in N$ with the required first $2n$ terms of their asymptotic expansions, namely

$$f(z) = -\cfrac{s_0}{z - a_0 - \cfrac{b_0^2}{z - a_1 - \cfrac{b_1^2}{z - a_2 - \cdots \cfrac{b_{n-2}^2}{z - a_{n-1} + f_n(z)}}}}.$$

Here $f_n(z) \in N$ and $f_n(z) = o(1)$.   Continuing the process to infinity, we obtain an infinite continued fraction.   This fraction possesses the characteristic property mentioned in Section **4.2** of Chapter 1 in relation to the power series we are discussing.   Therefore it coincides with the continued fraction which we introduced in Section **4.2** of Chapter 1.

It is worth noting that the operations we have performed are essentially nothing other than the consecutive divisions which were used already by Chebyshev and Markov to construct the continued fraction belonging to a given power series (the so-called associated continued fraction).

4. Canonical solution of the indeterminate Hamburger problem

4.1. We continue the study of the indeterminate moment problem, basing ourselves principally on Nevanlinna's formula

$$-\frac{A(z)\phi(z) - C(z)}{B(z)\phi(z) - D(z)} = \int_{-\infty}^{\infty} \frac{d\sigma(u)}{u - z}.$$

The solution of the moment problem $\sigma(u)$ is N-extremal if the function $\phi(z)$ corresponding to it in this formula is a real constant (finite or infinite). We denote this constant by τ and the solution† by $\sigma_\tau(u)$. The only points of increase of the function $\sigma_\tau(u)$ are the zeros $\lambda_k = \lambda_k(\tau)$ of the entire function

$$q(z) = \begin{cases} B(z)\tau - D(z) & \text{(if } \tau \neq \infty), \\ B(z) & \text{(if } \tau = \infty). \end{cases}$$

As regards the discontinuities

$$\mu_k = \mu_k(\tau) = \sigma_\tau(\lambda_k+0) - \sigma_\tau(\lambda_k-0)$$

of $\sigma_\tau(u)$, they may be found from the formula

$$\mu_k = \frac{A(\lambda_k)\tau - C(\lambda_k)}{B'(\lambda_k)\tau - D'(\lambda_k)} = \frac{1}{B'(\lambda_k)D(\lambda_k) - B(\lambda_k)D'(\lambda_k)}.$$

It follows from equations [1.23] and [1.15] that

$$B_n(\mu)D_n(\lambda) - B_n(\lambda)D_n(\mu) = b_{n-1}\{P_{n-1}(\lambda)P_n(\mu) - P_{n-1}(\mu)P_n(\lambda)\}.$$

Therefore

$$B_n'(\lambda)D_n(\lambda) - B_n(\lambda)D_n'(\lambda) = b_{n-1}\{P_n'(\lambda)P_{n-1}(\lambda) - P_n(\lambda)P_{n-1}'(\lambda)\}$$

$$= \sum_0^{n-1} [P_k(\lambda)]^2 = \frac{1}{\rho_{n-1}(\lambda)} \quad (-\infty < \lambda < \infty).$$

Thus

$$\mu_k = \lim_{n\to\infty} \rho_{n-1}(\lambda_k) = \rho(\lambda_k).$$

Thus the discontinuity of an N-extremal solution at each point of increase is equal to the maximum mass which may be concentrated at that point.

One has to add only very little to this result in order to prove the following proposition:

THEOREM 3.4.1. *Taking any point ξ $(-\infty < \xi < \infty)$, there exists one and only one solution $\sigma(u)$ of the full moment problem which has a concentrated mass at the point ξ equal to the maximum mass $\rho(\xi)$. This solution is N-extremal.*

Proof. In order to construct any solution of the required kind it is sufficient to find τ from the condition

$$B(\xi)\tau - D(\xi) = 0,$$

† This notation was introduced at the end of Section 4.2 in Chapter 2.

if $B(\xi) \neq 0$, and to put $\tau = \infty$ if $B(\xi) = 0$ and then to take the N-extremal solution $\sigma_\tau(u)$.

To prove that other solutions do not exist we put

$$s_k^* = s_k - \rho(\xi)\xi^k \quad (k = 0, 1, 2, \ldots).$$

The sequence $\{s_k^*\}_0^\infty$ is positive, since it permits the representation

$$s_k^* = \int_{-\infty}^\infty u^k \, d\sigma_\tau(u) - \rho(\xi)\xi^k = \sum_{\lambda_j \neq \xi} \rho(\lambda_j)\lambda_j^k \quad (k = 0, 1, 2, \ldots),$$

where $\lambda_j = \lambda_j(\tau)$ are the points of increase of the function $\sigma_\tau(u)$. Here the expressions given represent the unique solution of the auxiliary moment problem

$$s_k^* = \int_{-\infty}^\infty u^k \, d\sigma^*(u) \quad (k = 0, 1, 2, \ldots),$$

since, if this problem were indeterminate, it would have a solution with a positive mass at the point ξ and therefore the initial moment problem would have a solution with a mass at the point ξ exceeding $\rho(\xi)$.

Hence our statement follows.

DEFINITION. **3.4.1.** The N-extremal solutions of an indeterminate moment problem are called *canonical* solutions.

4.2. DEFINITION. **3.4.2.** We use the term *canonical solution of order* m for such a solution $\sigma(u)$ of an indeterminate moment problem to which there corresponds in Nevanlinna's formula a rational function $\phi(z) \in N$ of exactly m-th degree, which is real on the real axis.

Thus a canonical solution in the original sense is a canonical solution of order 0.

We agree to denote by V_m the aggregate of all canonical solutions of order m.

It follows directly from Nevanlinna's formula that a canonical solution of any order m has only discrete points of increase which are the zeros of an entire function of minimal exponential type.

For the following a certain interpolation problem is of importance:

In the half plane $\operatorname{Im} z > 0$ points $z_1, z_2, \ldots z_m$, are given, no two of which are equal. A certain indeterminate moment problem is considered, of which that solution $\sigma(u)$ is required for which

$$\int_{-\infty}^\infty \frac{d\sigma(u)}{u - z_k} = w_k \quad (k = 1, 2, \ldots, m), \qquad \ldots [3.25]$$

where the w_k $(k = 1, 2, \ldots, m)$ are given points in the half-plane $\operatorname{Im} w > 0$.

We refer here to necessary and sufficient conditions for the numbers w_k, which guarantee that the required function $\sigma(u)$ exists.

These conditions are easy to obtain with the aid of Nevanlinna's formula

$$-\frac{A(z)\phi(z)-C(z)}{B(z)\phi(z)-D(z)} = \int_{-\infty}^{\infty} \frac{d\sigma(u)}{u-z} \equiv w(z).$$

Replacing z in this formula by the numbers $z_1, z_2, \ldots, z_m$ and putting $w(z_k) = w_k$, we find the quantities $\phi(z_k) = \phi_k$. Now the problem reduces to the Nevanlinna-Pick interpolation problem with respect to the function $\phi(z) \in N$. According to Theorem 3.3.3 the required necessary and sufficient conditions consist in demanding that the hermitian form

$$\sum_{j,k=1}^{m} \frac{\phi_j - \bar{\phi}_k}{z_j - \bar{z}_k} \xi_j \bar{\xi}_k$$

should be non-negative; if this form is singular the function $\phi(z)$ is unique and in fact equal to a rational fraction of degree $< m$ which is real on the real axis (or to an infinite constant), and then $\sigma(u)$ is a canonical solution of order $< m$. In the following we shall write this condition in another form. Here we shall assume that there exists at least one solution $\sigma(u)$ for which conditions [3.25] are satisfied and we then take one further point z_{m+1}, different from any of those previously chosen. The question arises as to the nature of the set in the w plane traversed under these conditions by the point

$$w_{m+1} = \int_{-\infty}^{\infty} \frac{d\sigma(u)}{u-z_{m+1}}.$$

We denote this set by the symbol

$$K\begin{pmatrix} z_1 & z_2 & \cdots & z_m \\ w_1 & w_2 & \cdots & w_m \end{pmatrix}; \quad z_{m+1}\end{pmatrix}.$$

We assume to start with that there exists a non-denumerable set of solutions $\sigma(u)$ satisfying condition [3.25]. In this case it follows from what has been said that the set

$$K\begin{pmatrix} z_1 & z_2 & \cdots & z_m \\ w_1 & w_2 & \cdots & w_m \end{pmatrix}; \quad z_{m+1}\end{pmatrix}$$

is characterised by the fact that the form

$$\sum_{j,k=1}^{m} \frac{\phi_j - \bar{\phi}_k}{z_j - \bar{z}_k} \xi_j \bar{\xi}_k \qquad\qquad \ldots[3.26]$$

is positive while the form

$$\sum_{j,k=1}^{m+1} \frac{\phi_j - \bar{\phi}_k}{z_j - \bar{z}_k} \xi_j \bar{\xi}_k \qquad \ldots [3.26a]$$

is only non-negative, in other words by the fact that the form [3.26] is positive and that

$$\begin{vmatrix} \dfrac{\phi_1 - \bar{\phi}_1}{z_1 - \bar{z}_1} & \dfrac{\phi_1 - \bar{\phi}_2}{z_1 - \bar{z}_2} & \cdots & \dfrac{\phi_1 - \bar{\phi}_{m+1}}{z_1 - \bar{z}_{m+1}} \\ \cdots & & & \cdots \\ \dfrac{\phi_{m+1} - \bar{\phi}_1}{z_{m+1} - \bar{z}_1} & \dfrac{\phi_{m+1} - \bar{\phi}_2}{z_{m+1} - \bar{z}_2} & \cdots & \dfrac{\phi_{m+1} - \bar{\phi}_{m+1}}{z_{m+1} - \bar{z}_{m+1}} \end{vmatrix} \geqq 0.$$

This inequality may be rewritten in the form

$$A \,|\, \phi_{m+1} \,|^2 + B\phi_{m+1} + \bar{B}\bar{\phi}_{m+1} + C \geqq 0,$$

where

$$A = \begin{vmatrix} \dfrac{\phi_1 - \bar{\phi}_1}{z_1 - \bar{z}_1} & \cdots & \dfrac{\phi_1 - \bar{\phi}_m}{z_1 - \bar{z}_m} & \dfrac{1}{z_1 - \bar{z}_{m+1}} \\ \cdots & & & \\ \cdots & & & \\ \dfrac{\phi_m - \bar{\phi}_1}{z_m - \bar{z}_1} & \cdots & \dfrac{\phi_m - \bar{\phi}_m}{z_m - \bar{z}_m} & \dfrac{1}{z_m - \bar{z}_{m+1}} \\ \dfrac{1}{\bar{z}_1 - z_{m+1}} & \cdots & \dfrac{1}{\bar{z}_m - z_{m+1}} & 0 \end{vmatrix}.$$

Owing to the fact that the form [3.26] is positive, the quantity A is negative. Therefore the inequality obtained for ϕ_{m+1} shows that the point ϕ_{m+1} lies in the interior or on the boundary of a certain circle. The quantity w_{m+1} is connected with ϕ_{m+1} by a linear fractional transformation. Therefore the point w_{m+1} also lies within or on the boundary of a certain circle. Thus

$$K\begin{pmatrix} z_1 & z_2 \ldots z_m \\ w_1 & w_2 \ldots w_m \end{pmatrix};\ z_{m+1} \end{pmatrix}$$

is a circle.

We assume now that under the conditions [3.25] there exists only one solution $\sigma(u)$. In that case the value w_{m+1} of the integral

$$\int_{-\infty}^{\infty} \frac{d\sigma(u)}{u - z}$$

at the point $z = z_{m+1}$ is completely determined by its values at the previous points and therefore

$$K\begin{pmatrix} z_1 & z_2 \cdots z_m \\ w_1 & w_2 \cdots w_m \end{pmatrix}; \ z_{m+1}$$

is a point.

THEOREM **3.4.2.** *The solution $\sigma(u)$ of a moment problem is canonical of precise order m, if for any choice of numbers z_k, w_k, $(k = 1, 2, \ldots m+1)$* ($\mathrm{Im}\, z_k > 0$, $\mathrm{Im}\, w_k > 0$) *the point set*

$$K\begin{pmatrix} z_1 & z_2 \cdots z_m \\ w_1 & w_2 \cdots w_m \end{pmatrix}; \ z_{m+1}$$

is a circle and if the point w_{m+1} lies on its circumference.

Proof. It follows from the condition of the theorem that the form [*3.26*] is positive and the form [*3.26a*] non-negative and singular. Therefore by Theorem **3.3.3** there exists one and only one function $\phi(z) \in N$ satisfying the conditions

$$\phi(z_k) = \phi_k \quad (k = 1, 2, \ldots, m+1),$$

and this function is of the form

$$\phi(z) = \mu z + v + \sum_{l=1}^{n} \mu_l \frac{1 + u_l z}{u_l - z},$$

where $\mu \geqq 0$, $\mu_l > 0$, $v \geqq 0$; $u_l \geqq 0$. We must prove that the function $\phi(z)$ is exactly of degree m. The required proof is essentially contained in the discussion by means of which the last statement of **3.3.3.** was proved. However it is simplest to repeat the argument here. To do this we write down the given singular form for the values $\xi_k = \gamma_k$ for which it becomes zero:

$$0 = \sum_{j,k=1}^{m+1} \frac{\phi_j - \bar{\phi}_k}{z_j - \bar{z}_k} \gamma_j \bar{\gamma}_k = \mu \left| \sum_{j=1}^{m+1} \gamma_j \right|^2 + \sum_{l=1}^{n} \mu_l (1 + u_l^2) \left| \sum_{j=1}^{m+1} \frac{\gamma_j}{u_l - z_j} \right|^2$$

$$= \mu \left| \sum_{j=1}^{m+1} \gamma_j \right|^2 + \sum_{l=1}^{n} \mu_l (1 + u_l^2) \frac{|R_m(u_l)|^2}{\left| \prod_{j=1}^{m+1} (u_l - z_j) \right|^2},$$

where $R_m(u)$ is a polynomial of degree m, the leading coefficient of which is $\sum_{j=1}^{m+1} \gamma_j$. Since each term on the right hand side must be zero we must have

$$\mu \left| \sum_{j=1}^{m+1} \gamma_j \right|^2 = 0, \quad R_m(u_l) = 0 \quad (l = 1, 2, \ldots, n).$$

Therefore one of two statements must be true: either $\mu = 0$ and $n \leqq m$, or $\mu > 0$ and $n \leqq m-1$. In both cases $\phi(z)$ has degree $\leqq m$. However the degree of $\phi(z)$ cannot be less than m because this is incompatible with the positive property of the form [3.26]. Therefore $\phi(z)$ is exactly of degree m and the theorem is proved.

4.3. One can introduce a certain auxiliary and also indeterminate moment problem (which we shall denote by an asterisk) in such a way that the circle

$$K\begin{pmatrix} z_1 & z_2 & \cdots & z_m \\ w_1 & w_2 & \cdots & w_m \end{pmatrix}; \; z_{m+1}\end{pmatrix}$$

will be related directly to the usual circle $K_\infty^*(z_{m+1})$ of this auxiliary problem. Indeed we shall relate to each solution $\sigma(u)$ of our moment problem which satisfies condition [3.25] a function

$$\sigma^*(u) = \int_{-\infty}^u \frac{d\sigma(t)}{\displaystyle\prod_1^m |t - z_k|^2}.$$

Its moments

$$s_n^* = \int_{-\infty}^\infty u^n \, d\sigma^*(u) \quad (n = 0, 1, 2, \ldots) \qquad \ldots[3.27]$$

are uniquely determined by the moments s_i and the numbers z_k, w_k $(k = 1, 2, \ldots, m)$. Indeed it is always possible to write down the expansion

$$\frac{\lambda^n}{\displaystyle\prod_1^m |\lambda - z_k|^2} = \sum_{k=1}^m \left\{ \frac{A_k^{(n)}}{\lambda - z_k} + \frac{\bar{A}_k^{(n)}}{\lambda - \bar{z}_k} \right\} + R_{n-2m}(\lambda),$$

where $R_{n-2m}(\lambda)$ is a completely determined polynomial of degree $n-2m$, if $n \geqq 2m$ and $R_{n-2m}(\lambda) = 0$, if $n < 2m$. It follows from this expansion that

$$s_n^* = \sum_{k=1}^m \left\{ A_k^{(n)} w_k + \bar{A}_k^{(n)} \bar{w}_k \right\} + S\{R_{n-2m}(\lambda)\}. \qquad \ldots[3.28]$$

Conversely if the moments s_n^* $(n = 0, 1, 2, \ldots)$ are known, then the numbers $w_1, w_2, \ldots, w_n$ are uniquely determined and also all the original moments. For the first $2m$ relations [3.28] allow one to express the quantities† w_1, w_2, ... w_m in terms of the moments s_0^*, s_1^*, ... s_{2m-1}^*, and the remaining relations [3.28] permit one to express the quantity $\mathfrak{S}\{R_p(\lambda)\}$ $(p = 0, 1, 2, ...)$ in terms of the moments $s_0^*, s_1^*, ... s_{2m+p}^*$; by this means all the original moments are also expressed in terms of the moments $s_0^*, s_1^*, ... $. One can add to this that if $\sigma^*(u)$ is any solution of the problem [3.27], the function

$$\sigma(u) = \int_{-\infty}^{u} \prod_1^m |t - z_k|^2 \, d\sigma^*(t)$$

will be a solution of the original moment problem, and in fact a solution satisfying condition [3.25].

We take any point z_{m+1}. Then the circle

$$K\begin{pmatrix} z_1 & z_2 ... z_m \\ w_1 & w_2 ... w_m \end{pmatrix}; \; z_{m+1}$$

will correspond to the region covered by values of the integral

$$w^* = \int_{-\infty}^{\infty} \frac{d\sigma^*(u)}{u - z_{m+1}},$$

i.e. to the usual circle $K_\infty^*(z_{m+1})$ of the power moment problem [3.27]. It is easy to write down an explicit expression of this relation:

$$w_{m+1} = Aw^* + B.$$

We note simply that

$$A = \prod_{k=1}^m (z_{m+1} - z_k)(z_{m+1} - \bar{z}_k),$$

$$B = \sum_1^m (c_k w_k + d_k \bar{w}_k).$$

† The determinant

$$\begin{vmatrix} A_1^{(0)} & A_1^{(1)} ... A_1^{(2m-1)} \\ \bar{A}_1^{(0)} & \bar{A}_1^{(1)} ... \bar{A}_1^{(2m-1)} \\ \cdots \cdots \cdots \cdots \\ A_m^{(0)} & A_m^{(1)} ... A_m^{(2m-1)} \end{vmatrix}$$

is evidently non-zero since

$$A_k^{(n)} = \frac{z_k^n}{\phi(z_k)\,\phi'(z_k)}, \qquad \bar{A}_k^{(n)} = \frac{\bar{z}_k^n}{\bar{\phi}(\bar{z}_k)\,\bar{\phi}'(\bar{z}_k)},$$

where

$$\phi(\lambda) = \prod_1^m (\lambda - z_j), \quad \bar{\phi}(\lambda) = \prod_1^m (\lambda - \bar{z}_j).$$

Using the relation just established between the circles $K_\infty^*(z_{m+1})$ and

$$K\begin{pmatrix} z_1 & z_2 & \dots & z_m \\ w_1 & w_2 & \dots & w_m \end{pmatrix}; \quad z_{m+1}\begin{pmatrix} \end{pmatrix},$$

and also Theorem 2.3.3 we can formulate the following result.

THEOREM 3.4.3. *Assume that $\sigma(u)$ is a solution of an indeterminate moment problem and that $z_1, z_2, \dots z_m$ are some non-real (not necessarily different) numbers, and put*

$$\omega(u) = \int_{-\infty}^u \frac{d\sigma(t)}{\prod_1^m |t-z_k|^2}.$$

Then, for the set of all polynomials to be dense in L_ω^2, it is necessary and sufficient that $\sigma(u)$ should be a canonical solution of order $\le m$.

COROLLARY. 3.4.3. *A canonical solution of any order is V-extremal.*

Proof. Let $\sigma(u)$ be a canonical solution of order $\le m$. We introduce a function $\omega(u)$ in the same way as in Theorem 3.43. Assume further that $f(u)$ is an arbitrary function from L_σ^1. One can always find a function $g(u) \in L_\sigma^1$ and L_σ^2 such that

$$\int_{-\infty}^\infty |f(u)-g(u)|\, d\sigma(u) < \varepsilon,$$

where ε is an arbitrarily chosen positive number. Further, one can write down the inequality

$$\int_{-\infty}^\infty |g(u)-R(u)|\, d\sigma(u) \le \sqrt{\int_{-\infty}^\infty |g(u)-R(u)|^2\, d\omega(u)}$$
$$\times \sqrt{\int_{-\infty}^\infty \prod_1^m |u-z_k|^2\, d\sigma(u)},$$

where $R(u)$ is a polynomial. It then remains to apply Theorem 3.4.3.

4.4. We shall transform the condition for the interpolation problem of Section 4.2 to be soluble so that it is expressed directly in terms of the values w_k instead of in terms of the function $\phi(z)$. To do this we introduce the symbol.

$$\{w_\alpha, w_\beta\} = \frac{w_\alpha - \bar{w}_\beta}{z_\alpha - \bar{z}_\beta} - \sum_{k=0}^\infty \{P_k(z_\alpha)w_\alpha + Q_k(z_\alpha)\}\overline{\{P_k(z_\beta)w_\beta + Q_k(z_\beta)\}}.$$

It is easy to see the significance of this expression. For if $\sigma(u)$ is a solution of the moment problem for which

$$\int_{-\infty}^{\infty} \frac{d\sigma(u)}{u-z_\alpha} = w_\alpha, \quad \int_{-\infty}^{\infty} \frac{d\sigma(u)}{u-z_\beta} = w_\beta,$$

then

$$\frac{w_\alpha - \overline{w}_\beta}{z_\alpha - \overline{z}_\beta} = \int_{-\infty}^{\infty} \frac{d\sigma(u)}{(u-z_\alpha)(u-\overline{z}_\beta)}$$

is the scalar product of the functions

$$\frac{1}{u-z_\alpha}, \quad \frac{1}{u-z_\beta} \qquad \qquad \dots[3.29]$$

of L_σ^2. These functions possess generalized Fourier series

$$\frac{1}{u-z_\gamma} \sim \sum_{k=0}^{\infty} \{P_k(z_\gamma)w + Q_k(z_\gamma)\}P_k(u) \quad (\gamma = \alpha, \beta).$$

Therefore

$$\sum_{k=0}^{\infty} \{P_k(z_\alpha)w_\alpha + Q_k(z_\alpha)\}\overline{\{P_k(z_\beta)w_\beta + Q_k(z_\beta)\}}$$

is the scalar product in l^2 of those vectors which correspond to the functions [3.29]. The quantity $\{w_\alpha, w_\beta\}$ may thus be called the *Parseval defect* for the pair of functions [3.29].

Our statement is: *The form*

$$\sum_{\alpha,\beta=1}^{m} \frac{\phi_\alpha - \overline{\phi}_\beta}{z_\alpha - \overline{z}_\beta} \xi_\alpha \overline{\xi}_\beta \qquad \qquad \dots[3.27]$$

is equivalent to the form

$$\sum_{\alpha,\beta=1}^{m} \{w_\alpha, w_\beta\}\eta_\alpha \overline{\eta}_\beta. \qquad \qquad \dots[3.30]$$

Indeed, we have

$$\{w_\alpha, w_\beta\} = \lim_{n\to\infty} \left\{ \frac{w_\alpha - \overline{w}_\beta}{z_\alpha - \overline{z}_\beta} - \sum_{k=0}^{n-1} [P_k(z_\alpha)w_\alpha + Q_k(z_\alpha)]\overline{[P_k(z_\beta)w_\beta + Q_k(z_\beta)]} \right\}$$

$$= \lim_{n\to\infty} \left\{ \frac{w_\alpha - \overline{w}_\beta}{z_\alpha - \overline{z}_\beta} - w_\alpha \overline{w}_\beta I_1^{(n)} - w_\alpha I_2^{(n)} - \overline{w}_\beta I_3^{(n)} - I_4^{(n)} \right\}.$$

Applying formula [3.28] of Section 2.1 and the relation [1.23] we find that

$$I_1^{(n)} \equiv \sum_{k=0}^{n-1} P_k(z_\alpha)\overline{P_k(z_\beta)} = \frac{B_n(z_\alpha)\overline{D_n(z_\beta)} - D_n(z_\alpha)\overline{B_n(z_\beta)}}{z_\alpha - \bar{z}_\beta},$$

$$I_2^{(n)} \equiv \sum_{k=0}^{n-1} P_k(z_\alpha)\overline{Q_k(z_\beta)} = \frac{1 - [D_n(z_\alpha)\overline{A_n(z_\beta)} - B_n(z_\alpha)\overline{C_n(z_\beta)}]}{z_\alpha - \bar{z}_\beta},$$

$$I_3^{(n)} \equiv \sum_{k=0}^{n-1} Q_k(z_\alpha)\overline{P_k(z_\beta)} = \frac{-1 + [A_n(z_\alpha)\overline{D_n(z_\beta)} - C_n(z_\alpha)\overline{B_n(z_\beta)}]}{z_\alpha - \bar{z}_\beta},$$

$$I_4^{(n)} \equiv \sum_{k=0}^{n-1} Q_k(z_\alpha)\overline{Q_k(z_\beta)} = \frac{A_n(z_\alpha)\overline{C_n(z_\beta)} - C_n(z_\alpha)\overline{A_n(z_\beta)}}{z_\alpha - \bar{z}_\beta}.$$

Therefore

$$\{w_\alpha, w_\beta\} = \lim_{n \to \infty} \frac{K_n}{z_\alpha - \bar{z}_\beta} = \frac{K}{z_\alpha - \bar{z}_\beta},$$

where

$$K_n = [D_n(z_\alpha)w_\alpha + C_n(z_\alpha)]\overline{[B_n(z_\beta)w_\beta + A_n(z_\beta)]}$$
$$- [B_n(z_\alpha)w_\alpha + A_n(z_\alpha)]\overline{[D_n(z_\beta)w_\beta + C_n(z_\beta)]}$$

and

$$K = [D(z_\alpha)w_\alpha + C(z_\alpha)]\overline{[B(z_\beta)w_\beta + A(z_\beta)]}$$
$$- [B(z_\alpha)w_\alpha + A(z_\alpha)]\overline{[D(z_\beta)w_\beta + C(z_\beta)]}.$$

Also, since

$$\phi_\alpha = \frac{D(z_\alpha)w_\alpha + C(z_\alpha)}{B(z_\alpha)w_\alpha + A(z_\alpha)},$$

we have

$$\{w_\alpha, w_\beta\} = [B(z_\alpha)w_\alpha + A(z_\alpha)]\overline{[B(z_\beta)w_\beta + A(z_\beta)]} \frac{\phi_\alpha - \bar{\phi}_\beta}{z_\alpha - \bar{z}_\beta},$$

and therefore the form [3.26] goes over into [3.30] if we put

$$\eta_\alpha = \frac{\xi_\alpha}{B(z_\alpha)w_\alpha + A(z_\alpha)} \quad (\alpha = 1, 2, ..., m).$$

Let us now formulate the result we have obtained.

THEOREM 3.4.4.† *For a solution of the power moment problem to exist that satisfies the conditions*

$$\int_{-\infty}^{\infty} \frac{d\sigma(u)}{u - z_k} = w_k \quad (\text{Im } z_k > 0; \; k = 1, 2, \ldots, n \leqq \infty),$$

it is necessary and sufficient that for any finite $m \leqq n$ the following form should be non-negative:

$$\sum_{\alpha, \beta = 1}^{m} \{w_\alpha, w_\beta\} \eta_\alpha \bar{\eta}_\beta.$$

Addenda and Problems

1. The Stieltjes-Perron inversion formula.

Put

$$\phi(\xi, \eta) = \int_{-\infty}^{\infty} \left\{ \frac{\eta}{(\xi - u)^2 + \eta^2} - \frac{\eta}{1 + u^2} \right\} d\psi(u) \quad (-\infty < \xi < \infty, \eta > 0),$$

where $\psi(u)$ has bounded variation in every finite interval and

$$\int_{-\infty}^{\infty} \frac{|d\psi(u)|}{1 + |u|^3} < \infty.$$

In that case we have, for any real t and c

$$\frac{\psi(t+0) + \psi(t-0)}{2} - \frac{\psi(c+0) + \psi(c-0)}{2} = \lim_{\eta \to 0} \frac{1}{\pi} \int_c^t \phi(\xi, \eta) \, d\xi.$$

Proof. We take for definiteness $c < t$ and include the segment $[c, t]$ in the interval $(-A, B)$ the ends of which are points of continuity of $\psi(u)$. Then we put

$$\delta(\eta) = \frac{1}{\pi} \int_c^t \phi(\xi, \eta) \, d\xi - \frac{1}{\pi} \int_c^t d\xi \int_{-A}^B \frac{\eta}{(\xi - u)^2 + \eta^2} \, d\psi(u).$$

Then on the one hand we have

$$|\delta(\eta)| \leqq \frac{1}{\pi} \int_c^t d\xi \int_{-A}^B \frac{\eta}{1 + u^2} |d\psi(u)|$$
$$+ \frac{\eta}{\pi} \int_c^t d\xi \left\{ \int_{-\infty}^{-A} \left| \frac{1}{(\xi - u)^2 + \eta^2} - \frac{1}{1 + u^2} \right| |d\psi(u)| \right.$$
$$\left. + \int_B^\infty \left| \frac{1}{(\xi - u)^2 + \eta^2} - \frac{1}{1 + u^2} \right| |d\psi(u)| \right\},$$

† See AKHIEZER[5].

from which it is evident that $\delta(\eta) = o(1)$ for $\eta \to 0$ and on the other hand $\delta(\eta)$ can be represented in the form

$$\delta(\eta) = \frac{1}{\pi} \int_c^t \phi(\xi, \eta) \, d\xi - \frac{1}{\pi} \int_{-A}^B \left\{ \arctan \frac{t-u}{\eta} - \arctan \frac{c-u}{\eta} \right\} d\psi(u),$$

whence, integrating by parts we find that

$$\delta(\eta) = \frac{1}{\pi} \int_c^t \phi(\xi, \eta) \, \delta\xi - \frac{1}{\pi} \int_{-A}^B \left\{ \frac{\eta}{(u-t)^2 + \eta^2} - \frac{\eta}{(u-c)^2 + \eta^2} \right\} \psi(u) \, du$$

$$+ \frac{1}{\pi} \psi(B) \left[\arctan \frac{B-t}{\eta} - \arctan \frac{B-c}{\eta} \right]$$

$$+ \frac{1}{\pi} \psi(-A) \left[\arctan \frac{A+t}{\eta} - \arctan \frac{A+c}{\eta} \right].$$

Therefore for $\eta \to 0$ we have

$$\frac{1}{\pi} \int_c^t \phi(\xi, \eta) \, d\xi - \frac{1}{\pi} \int_{-A}^B \left\{ \frac{\eta}{(u-t)^2 + \eta^2} - \frac{\eta}{(u-c)^2 + \eta^2} \right\} \psi(u) \, du = o(1),$$

and to obtain the required result it remains only to use the well known properties of the Poisson integral.

We note two particular cases

α. Let

$$f(z) = \int_{-\infty}^\infty \frac{d\omega(u)}{u-z} \quad (\text{Im } z > 0),$$

where $\omega(u)$ is a real function, of bounded variation in every finite interval, such that

$$\int_{-\infty}^\infty \frac{|d\omega(u)|}{1+|u|} < \infty.$$

In this case

$$\text{Im } f(\xi + i\eta) = \int_{-\infty}^\infty \frac{\eta}{(\xi-u)^2 + \eta^2} \, d\omega(u),$$

and therefore

$$\frac{\omega(t+0) + \omega(t-0)}{2} - \frac{\omega(c+0) + \omega(c-0)}{2} = \lim_{\eta \to 0} \frac{1}{\pi} \int_c^t \text{Im } f(\xi + i\eta) \, d\xi.$$

This was just the formula given by STIELTJES[3] and PERRON[1].

β. Assume that $f(z)$ is an arbitrary function of class N, so that

$$f(z) = \mu z + v + \int_{-\infty}^{\infty} \frac{1+uz}{u-z} \, d\sigma(u) \quad (\text{Im } z > 0),$$

where $\mu \geqq 0$, $v \leqq 0$, and $\sigma(u)$ is a non-decreasing function of bounded variation.

In this case

$$\text{Im } f(\xi + i\eta) - \mu\eta = \int_{-\infty}^{\infty} \frac{\eta(1+u^2)}{(\xi-u)^2 + \eta^2} \, d\sigma(u),$$

and therefore

$$\frac{\psi(t+0) + \psi(t-0)}{2} - \frac{\psi(c+0) + \psi(c-0)}{2} = \lim_{\eta \to 0} \frac{1}{\pi} \int_c^t \text{Im } f(\xi + i\eta) \, d\xi,$$

where

$$\psi(u) = \int_0^u (1+\lambda^2) \, d\sigma(\lambda).$$

2. In order that a function $f(z)$ of class N should be continuous and real on the negative half of the real axis it is necessary and sufficient that the class N should also contain the function

$$\frac{z}{1+z^2}\{f(z) - \text{Re } f(i) - z \,\text{Im } f(i)\} = g(z)$$

(AKHIEZER and KREIN[2]).

Hint. Since $f(z) \in N$ we have

$$f(z) = \mu z + v + \int_{-\infty}^{\infty} \frac{1+uz}{u-z} \, d\sigma(u),$$

and therefore

$$\frac{z}{1+z^2}\{f(z) - \text{Re } f(i) - z \,\text{Im } f(i)\} = \int_{-\infty}^{\infty} \frac{u \, d\sigma(u)}{u-z} - \int_{-\infty}^{\infty} d\sigma(u),$$

whence

$$g(z) + \int_{-\infty}^{\infty} d\sigma(u) = \int_{-\infty}^{\infty} \frac{d\psi(u)}{u-z},$$

where

$$\psi(u) = \int_0^u t \, d\sigma(t).$$

It then only remains to apply the Stieltjes-Perron inversion formula.

3. The function $f(z)$ permits the representation

$$f(z) = \alpha + \int_0^\infty \frac{d\tau(u)}{u-z} \quad (0 < \arg z < 2\pi),$$

where $\alpha \geqq 0$ and $\tau(u)$ is a non-decreasing function and

$$\int_0^\infty \frac{d\tau(u)}{1+u} < \infty,$$

if and only if

α) $f(z) \in N$ and

β) The function $f(z)$ is continuous and positive on the negative half of the real axis. (This theorem is due to Krein. Conditions α) and β) are essentially the same as those in the well-known lemma due to Löwner).

The only thing that requires a proof is that the properties α) and β) are sufficient to ensure the required representation of the function $f(z)$. We find as consequences of these properties that

$$f(z) = \mu z + \nu + \int_0^\infty \frac{1+uz}{u-z} \, d\sigma(u) \quad (\mu \geqq 0)$$

and that for any $c > 0$

$$-\mu c + \nu + \int_0^\infty \frac{1-cu}{u+c} \, d\sigma(u) > 0,$$

whence

$$\int_0^T \frac{cu}{u+c} \, d\sigma(u) < \nu - \mu c + \int_0^\infty \frac{d\sigma(u)}{u+c},$$

where $T > 0$ is arbitrary. Therefore $\mu = 0$ and

$$\int_0^\infty u \, d\sigma(u) \leqq \nu.$$

It then only remains to put

$$\alpha = \nu - \int_0^\infty u \, d\sigma(u), \quad \tau(u) = \int_0^u (1+t^2) \, d\sigma(t).$$

4. In order that the function $f(z)$ of class N should be continuous and positive on the negative half of the real axis it is necessary and sufficient that the function $zf(z)$ should also belong to the class N.

Proof. The necessity follows from the theorem of example **3.** In order to prove sufficiency, we write down the integral representation of the function $zf(z) \in N$:

$$zf(z) = \alpha z + \beta + \int_{-\infty}^{\infty} \frac{1+uz}{u-z} \, d\omega(u). \qquad \dots[3.31]$$

Here $\alpha \geqq 0$, $\beta \gtrless 0$ and $\omega(u)$ is a non-decreasing function of bounded variation. Further, since also $f(z) \in N$ the analogous representation is valid:

$$f(z) = \alpha_1 z + \beta_1 + \int_{-\infty}^{\infty} \frac{1+uz}{u-z} \, d\omega_1(u).$$

Hence it follows that

$$zf(z) = \alpha_1 z^2 + \beta_1 z - \int_{-\infty}^{\infty} d\omega_1(u) + \int_{-\infty}^{\infty} \left[\frac{1+uz}{u-z} - z \right] u \, d\omega_1(u). \quad \dots[3.32]$$

Subtracting [3.32] from [3.31] we obtain

$$0 = \alpha z + \beta - \alpha_1 z^2 - \beta_1 z + \int_{-\infty}^{\infty} d\omega_1(u) + z \int_{-\infty}^{\infty} d\omega(u)$$

$$+ \int_{-\infty}^{\infty} \left[\frac{1+uz}{u-z} - z \right] [d\omega(u) - u \, d\omega_1(u)].$$

Putting

$$\psi(u) = \int_{0}^{u} (1+\lambda^2) \, d\omega(\lambda) - \int_{0}^{u} \lambda(1+\lambda^2) \, d\omega_1(\lambda)$$

and splitting off the imaginary part, we find that

$$2\alpha_1 \xi \eta + \gamma \eta = \int_{-\infty}^{\infty} \left\{ \frac{\eta}{(\xi-u)^2 + \eta^2} - \frac{\eta}{1+u^2} \right\} d\psi(u),$$

where $\gamma \gtrless 0$. Using the general inversion formula of example **1** we conclude that

$$\frac{\psi(u+0) + \psi(u-0)}{2} = \text{const.}$$

Therefore $d\omega(u) = u \, d\omega_1(u)$ and $d\omega_1(u) = 0$ for $u < 0$, and thus it follows from [3.31] that

$$f(z) = \alpha + \frac{\beta}{z} + \int_{0}^{\infty} \left\{ \frac{1}{z} + \frac{1+u^2}{u-z} \right\} d\omega_1(u) = \alpha + \int_{0}^{\infty} \frac{d\tau(u)}{u-z},$$

where $\tau(u)$ can only be a non-decreasing function.

5. A function $f(z)$ (Im $z > 0$) permits the representation

$$f(z) = z\left\{a + \int_0^\infty \frac{d\tau(u)}{u - z^2}\right\},$$

where the constant $a \geqq 0$ and the non-decreasing function $\tau(u)$ satisfies the condition

$$\int_0^\infty \frac{d\tau(u)}{1 + u} < \infty,$$

if and only if $f(z)$ belongs to the class N and takes on purely imaginary values on the imaginary axis (Krein).

6. The function $f(z) \in N$ satisfies the inequality

$$\int_1^\infty \frac{\operatorname{Im} f(iy)}{y^\alpha} \, dy < \infty,$$

where α is a number in the interval $(0, 2)$, if and only if $f(z)$ permits the representation

$$f(z) = \beta + \int_{-\infty}^\infty \left\{\frac{1}{u - z} - \frac{u}{1 + u^2}\right\} d\tau(u), \qquad \dots [3.32]$$

where the constant β is real and the non-decreasing function $\tau(u)$ satisfies the condition

$$\int_{-\infty}^\infty \frac{d\tau(u)}{1 + |u|^\alpha} < \infty.$$

If $\alpha \leqq 1$ the representation $[3.32]$ can be written in the form

$$f(z) = \gamma + \int_{-\infty}^\infty \frac{d\tau(u)}{u - z} \qquad (\gamma \gtrless 0)$$

(KATZ[1]).

7. If $f(z) \in N$ then for Im $z > 0$ we have

$$\frac{\operatorname{Im} z}{1 + |z|^2} < \frac{\operatorname{Im} f(z)}{\operatorname{Im} f(i)} < \frac{1 + |z|^2}{\operatorname{Im} z}.$$

Hint Use the inequality

$$\left|\frac{f(z) - f(i)}{f(z) - \overline{f(i)}}\right| \leqq \left|\frac{z - i}{z + i}\right|,$$

which expresses Schwartz' lemma. It follows from this inequality that the point $f(z)$ lies in the strip

$$\frac{1}{H} \operatorname{Im} f(i) \leqq \operatorname{Im} f(z) \leqq H \operatorname{Im} f(i),$$

where H is determined from the relation

$$\frac{H-1}{H+1} = \left|\frac{z-i}{z+i}\right|.$$

8. Assume that $f(z)$ ranges over the set $\mathfrak{M} \subset \mathbb{N}$, determined as follows:

a) All functions $f(z) \in \mathfrak{M}$ have a fixed value w_0 at a given point z_0 of the upper half-plane;

b) In a given interval (α, β) of the real axis every function $f(z) \in \mathfrak{M}$ is continuous and real.

It is required to determine the domain $\mathfrak{D}$ in the w-plane occupied by those values which the functions of the set $\mathfrak{M}$ assume at a given point $z = \zeta\ (\neq z_0)$ of the upper half-plane.

If condition b) were not imposed then by the Schwartz lemma the required domain would be the Lobachevski "circle" K with "centre" w_0 and "radius" equal to the non-Euclidian distance from the point ζ to the point z_0. The equation of the circumference of this circle is the following:

$$\left|\frac{w-w_0}{w-\bar{w}_0}\right| = \left|\frac{\zeta-z_0}{\zeta-\bar{z}_0}\right|, \qquad \dots[3.33]$$

and it may be represented in parametric form as

$$w = u_0 + v_0\frac{y_0-tx_0+t\zeta}{ty_0+x_0-\zeta}, \qquad \dots[3.33a]$$

where $u_0 + iv_0 = w_0$, $x_0 + iy_0 = z_0$ and t ranges over the real axis.

With the condition b) the domain $\mathfrak{D}$ will be a part of the circle K. If $\alpha = -\infty$, $\beta = \infty$, the set $\mathfrak{M}$ consists of only the single function

$$f(z) = \frac{v_0}{y_0}(z-x_0)+u_0,$$

and therefore the domain $\mathfrak{D}$ degenerates into the point w_0. Disregarding this case, we assume that $-\infty \leqq \alpha < \beta$. Now the domain $\mathfrak{D}$ will contain the point

$$w_\infty = \frac{v_0}{y_0}(\zeta-x_0)+u_0$$

and will consist of the segment of circle bounded by the circumference [3.33] and the chord

$$w = \theta w_\alpha + (1-\theta)w_\beta \quad (0 \leqq \theta \leqq 1),$$

where the points w_α and w_β are determined by equation [3.33a] for $t = (\alpha - x_0)/y_0$ and $t = (\beta - x_0)/y_0$ respectively. (AKHIEZER and KREIN[1]).

9. Assume an indeterminate moment problem

$$s_k = \int_{-\infty}^{\infty} u^k \, d\sigma(u) \quad (k = 0, 1, 2, ...) \qquad ...[3.34]$$

to be given and assume that W_0 is the closed convex hull of its canonical solutions of zero order.

Prove that $\sigma(u) \in W_0$ if and only if the function $\phi(z)$ in Nevanlinna's formula

$$\int_{-\infty}^{\infty} \frac{d\sigma(u)}{u - z} = - \frac{A(z)\phi(z) - C(z)}{B(z)\phi(z) - D(z)}$$

is equal to

$$\phi(z) = \frac{D(z)}{B(z)} + \frac{1}{\displaystyle\int_{-\infty}^{\infty} \frac{d\mu(\tau)}{\tau - \dfrac{D(z)}{B(z)}}}, \qquad ...[3.35]$$

where $u(\tau)$ is a non-decreasing function, the total variation of which is ≤ 1. A solution, $\sigma(u)$ of the moment problem corresponding to the function [3.35] can be represented in the form

$$\sigma(u) = \left[1 - \int_{-\infty}^{\infty} d\mu(\tau) \right] \sigma_\infty(u) + \int_{-\infty}^{\infty} \sigma_\tau(u) \, d\mu(\tau),$$

where $\sigma_\tau(u)$ is that canonical solution of problem [3.34] for which $\phi(z) \equiv \tau$ (GLAZMAN[1]).

10. The set V of all solutions of Hamburger's indefinite moment problem

$$s_k = \int_{-\infty}^{\infty} u^k \, d\sigma(u) \quad (k = 0, 1, 2, ...) \qquad ...[3.36]$$

is essentially closed. This means that if $\{\sigma^{(n)}(u)\}_1^\infty$ is any sequence of solutions and $\sigma(u)$ is some function at all points of continuity of which

$$\sigma(u) = \lim_{n \to \infty} \sigma^{(n)}(u),$$

then $\sigma(u)$ is also a solution of problem [3.36].

Prove that the totality of all canonical solutions (in the narrow and extended senses) forms a set which is everywhere dense in V. Hence it follows that the set V is essentially the closure of the totality of its limiting points (GLAZMAN and NAIMAN[1]).

11. If a Nevanlinna matrix is not generated by an indeterminate moment problem the elements of this matrix need not be functions of minimal exponential type. The question arises naturally how strong the consequences are for a quartet of transcendental functions of the fact alone that they form a Nevanlinna matrix. This question was answered by KREIN[10], with the aid of the following general proposition on entire functions which he himself had established immediately previously (see KREIN[7, 10]): If the zeros of an entire function $g(z)$ are real† and if it permits the absolutely convergent expansion

$$\frac{1}{g(z)} = \sum_1^\infty \frac{A_k}{z-h_k} \qquad \qquad ...[3.37]$$

(which means that

$$\sum_1^\infty \frac{|A_k|}{1+|h_k|} < \infty),$$

then $g(z)$ is an entire function of finite degree and

$$\int_{-\infty}^\infty \frac{\ln^+ |g(x)|}{1+x^2} dx < \infty. \qquad \qquad ...[3.38]$$

The proposition just formulated has one consequence which will be required in the next example. This is: if $f(z)$ is an entire function with real zeros and if the absolutely convergent expansion

$$\frac{1}{f(z)} = \frac{A}{z} + z\sum_1^\infty \frac{A_k}{(z-\alpha_k)\alpha_k} + B + Cz, \qquad ...[3.39]$$

is valid, then $C = 0$.

---

† As was shown by Krein the theorem is true also in the case that one postulates in place of real zeros the convergence of the series

$$\sum_1^\infty \left| \operatorname{Im} \frac{1}{h_k} \right|.$$

The full proof for the general case may be found in the monograph by LEVIN[1].
  Very essential generalizations of Krein's theorem are due to OSTROVSKI[1] and to MATSAEV[1]. In particular Matsaev proved that an entire function $g(z)$ is of exponential type and satisfies condition [3.38] even in the case when the inequality

$$|g(z)| \geq \frac{1}{C} \exp\left\{ -Cr^\alpha \left(1+\frac{1}{|y|^\beta}\right)\right\}$$

holds everywhere in the plane where

$$r=|z|, \quad y=\operatorname{Im} z, \quad 0 \leq \alpha < 1, \quad \beta \geq 0, \quad C>0.$$

Considering the expansion of the function

$$\frac{1}{f(z)(z-\beta)(z-\gamma)},$$

where the numbers β and γ are real and different from the zeros of $f(z)$ we find by Krein's theorem that $f(z)(z-\beta)(z-\gamma)$, and therefore also $f(z)$ itself, is an entire function of exponential type. On the other hand if $C \neq 0$, then by virtue of the expansion [3.39] the product $zf(z)$ is bounded within the angles

$$\delta \leqq \arg z \leqq \pi - \delta, \quad \pi + \delta \leqq \arg z \leqq 2\pi - \delta \quad \left(0 < \delta < \frac{\pi}{2}\right).$$

By the Phragmén-Lindelöf theorem this product will be bounded also within the supplementary angles, which is absurd. Thus C must be zero.

12. For a Nevanlinna matrix to exist, one element of which is the entire transcendental function $f(z)$, it is necessary and sufficient that $f(z)$ should be real, should have only real zeros and should permit the absolutely convergent expansion

$$\frac{1}{f(z)} = \frac{A}{z} + z \sum_{1}^{\infty} \frac{A_k}{(z-\alpha_k)\alpha_k} + B \qquad \ldots [3.40]$$

(KREIN[10]).

By the theorem of the previous example a function $f(z)$ which satisfies the conditions of the present theorem is of exponential type and for it the integral

$$\int_{-\infty}^{\infty} \frac{\ln^+ |f(x)|}{1+x^2} dx$$

converges.

It is easy to see that if

$$\left\| \begin{array}{cc} a(z) & c(z) \\ b(z) & d(z) \end{array} \right\| \qquad \ldots [3.41]$$

is a Nevanlinna matrix then

$$\left\| \begin{array}{cc} -b(z) & -d(z) \\ a(z) & c(z) \end{array} \right\|$$

will also be a Nevanlinna matrix. Therefore one can take it that the function $f(z)$ is one of the " denominators " of a Nevanlinna matrix, say

$$f(z) = b(z).$$

In order to prove sufficiency, we assume that there exists the absolutely convergent expansion

$$\frac{1}{b(z)} = \frac{A_0}{z} + z\sum_1^\infty \frac{A_k}{(z-\alpha_k)\alpha_k} + K.$$

It permits us to introduce functions $a(z)$ and $d(z)$ according to the formulae

$$\frac{a(z)}{b(z)} = \frac{m_0\,|\,A_0\,|}{z} + z\sum_1^\infty \frac{m_k\,|\,A_k\,|}{(z-\alpha_k)\alpha_k} - \alpha z + \beta, \qquad \ldots[3.42]$$

$$\frac{d(z)}{b(z)} = \frac{|\,A_0\,|}{m_0 z} + z\sum_1^\infty \frac{|\,A_k\,|}{m_k(z-\alpha_k)\alpha_k} - \gamma z + \delta, \qquad \ldots[3.43]$$

where β and δ are arbitrary real numbers, $\alpha \geqq 0$, $\gamma \geqq 0$, $m_0 > 0$, $m_k > 0$ and

$$\sum_1^\infty \frac{|\,A_k\,|}{\alpha_k^2}\left(m_k + \frac{1}{m_k}\right) < \infty.$$

(for instance one can put $m_k = 1$). As the function $c(z)$ we must then take

$$c(z) = \frac{a(z)\,d(z)}{b(z)} - \frac{1}{b(z)},$$

and evaluation of residues shows that this is an entire function.

One simply has to verify that the function

$$w = -\frac{a(z)t - c(z)}{b(z)t - d(z)}$$

belongs to the class N for any real t. To do this we represent w in the form

$$w = -\frac{a(z)}{b(z)} - \frac{\dfrac{1}{b^2(z)}}{t - \dfrac{d(z)}{b(z)}} = -\frac{a(z)}{b(z)} - \frac{\dfrac{1}{b^2(z)}}{\tau - i\,\mathrm{Im}\,\dfrac{d(z)}{b(z)}},$$

where τ is real. Putting

$$-\mathrm{Im}\,\frac{a(z)}{b(z)} = A, \quad -\mathrm{Im}\,\frac{d(z)}{b(z)} = C, \quad \frac{1}{b(z)} = \xi + iB,$$

where ξ and B are real, we find that

$$\text{Im } w = A - \frac{2\tau\xi B + (B^2 - \xi^2)C}{\tau^2 + C^2}$$

$$= \frac{1}{\tau^2 + C^2}\left\{\left(\sqrt{A}\tau - \xi\frac{B}{\sqrt{A}}\right)^2 + C(AC - B^2) + \frac{\xi^2}{A}(AC - B^2)\right\}.$$

It remains to bear in mind that for Im $z > 0$ we have, by virtue of the expansions [3.42] and [3.43], that

$$A > 0, \quad C > 0,$$

and by the Cauchy inequality

$$B^2 = \left[\text{Im }\frac{1}{b(z)}\right]^2 \leqq \text{Im }\frac{a(z)}{b(z)} \cdot \text{Im }\frac{d(z)}{b(z)} = AC.$$

Thus the proof of sufficiency is complete.

We now prove necessity. To do this we establish that each of the functions

$$\Phi_1(z) = -\frac{a(z)}{b(z)} + 2\frac{1}{b(z)} - \frac{d(z)}{b(z)},$$

$$\Phi_2(z) = -\frac{a(z)}{b(z)} - 2\frac{1}{b(z)} - \frac{d(z)}{b(z)},$$

$\quad\quad\quad\quad\quad\quad\quad\quad\quad\quad\quad\quad\quad$ ...[3.44]

generated by the Nevanlinna matrix belongs to the class N. Since these functions are meromorphic, their integral representations will have the form of absolutely convergent expansions:

$$\Phi_1(z) = \frac{A'}{z} + z\sum_1^\infty \frac{A_k'}{(z - \alpha_k)\alpha_k} + B' + C'z,$$

$$\Phi_2(z) = \frac{A''}{z} + z\sum_1^\infty \frac{A_k''}{(z - \alpha_k)\alpha_k} + B'' + C''z.$$

Hence we obtain by subtraction the absolutely convergent expansion

$$\frac{1}{b(z)} = \frac{A}{z} + z\sum_1^\infty \frac{A_k}{(z - \alpha_k)\alpha_k} + B + Cz.$$

As a result of the corollary in the previous example we have $C = 0$ and this concludes the proof of necessity.

The verification of the fact that the functions [3.44] belong to the class N can proceed as follows:

We take a real point

$$t_0 = \frac{\operatorname{Im} d(z)}{\operatorname{Im} b(z)} \quad (\operatorname{Im} z > 0),$$

which, by use of the formula

$$w = -\frac{a(z)t - c(z)}{b(z)t - d(z)} \qquad \qquad ...[3.45]$$

must go over into the point w_0 of the half-plane $\operatorname{Im} w > 0$. By simple calculation we find that

$$\operatorname{Im} w_0 = -\operatorname{Im} \frac{a(z)}{b(z)} + \frac{\left[\operatorname{Im} \dfrac{1}{b(z)}\right]^2}{\operatorname{Im} \dfrac{d(z)}{b(z)}}.$$

Now the problem reduces to proving the inequality

$$\operatorname{Im} \frac{d(z)}{b(z)} < 0 \quad (\operatorname{Im} z > 0). \qquad \qquad ...[3.46]$$

Indeed it will follow from this that

$$\operatorname{Im} \frac{a(z)}{b(z)} \operatorname{Im} \frac{d(z)}{b(z)} - \left[\operatorname{Im} \frac{1}{b(z)}\right]^2 > 0 \quad (\operatorname{Im} z > 0)$$

and therefore

$$\operatorname{Im} \left\{ -\frac{a(z)}{b(z)} \pm 2 \frac{1}{b(z)} - \frac{d(z)}{b(z)} \right\}$$

$$\geqq \operatorname{Im} \left\{ -\frac{a(z)}{b(z)} \right\} - 2 \left| \operatorname{Im} \frac{1}{b(z)} \right| + \operatorname{Im} \left\{ -\frac{d(z)}{b(z)} \right\}$$

$$> \left\{ \sqrt{\operatorname{Im} \left[-\frac{a(z)}{b(z)} \right]} - \sqrt{\operatorname{Im} \left[-\frac{d(z)}{b(z)} \right]} \right\}^2 \geqq 0.$$

In order to prove the inequality [3.46] we use the fact that for fixed z ($\operatorname{Im} z > 0$) the function [3.45] transforms the real axis of the t plane into a certain circle $k(z)$ lying in the half plane $\operatorname{Im} w > 0$. We can state that in this transformation the half-plane $\operatorname{Im} t > 0$ goes over into the interior of the circle $k(z)$. Indeed, if one assumes the opposite this implies that as t varies from $-\infty$ to ∞ the point w goes round the circumference of $k(z)$ in a clockwise sense. But then, by letting the imaginary part of z tend to zero we find that in the limit when z becomes real and the circle $k(z)$ goes over into the real axis of the w plane the variation of t from $-\infty$ to ∞ will correspond to a motion along the real axis in the w plane in the negative direction, which is impossible since for real z the half-plane Im $t > 0$ goes over into the half-plane Im $w > 0$, by virtue of the fact that the elements of the Nevanlinna matrix are real and that $a(z)d(z) - b(z)c(z) = 1$.

We assume now that [3.46] does not hold and therefore that at some point z of the half-plane Im $z > 0$ the inequality

$$\text{Im} \frac{d(z)}{b(z)} \geqq 0$$

holds. In this case the point

$$t_1 = \frac{d(z)}{b(z)}$$

lies in the half-plane Im $t > 0$ and by the foregoing this point must go over into a point w_1 lying within the half-plane Im $w > 0$ while at the same time

$$w_1 = -\frac{a(z)t_1 - c(z)}{b(z)t_1 - d(z)} = \infty.$$

Chapter 4

INCLUSION OF THE POWER MOMENT PROBLEM
IN THE SPECTRAL THEORY OF OPERATORS

In this chapter we shall discuss the operator approach to the moment problem and shall show how its basic ideas may be derived by means of spectral theory. However, the connection between the moment problem and spectral theory is not confined to this and the advantage brought about by this relationship is mutual since much of what is done in the moment problem can serve as a starting point and a model for more general constructions in the theory of operators. The discussion of this side of the question is extremely interesting and instructive but goes beyond the scope of this book.

In the present chapter it is assumed that the reader is familiar with the elements of operator theory.† Since the theory of operators lacks a unified terminology and a unified notation we have considered it necessary to give some definitions and recall some facts from that subject.

1. The operator approach to the moment problem

1.1 We shall take as our starting point the $\mathscr{J}$-matrix

$$\left\| \begin{array}{cccccc} a_0 & b_0 & 0 & 0 & 0 & \dots \\ b_0 & a_1 & b_1 & 0 & 0 & \dots \\ 0 & b_1 & a_2 & b_2 & 0 & \dots \\ \cdot & \cdot & \cdot & \cdot & \cdot & \cdot & \cdot & \cdot & \cdot & \cdot & \cdot \\ \cdot & \cdot & \cdot & \cdot & \cdot & \cdot & \cdot & \cdot & \cdot & \cdot & \cdot \end{array} \right\|$$

and shall consider it as the matrix of a certain linear operator A in a separable Hilbert space H. For this purpose we take in H an orthonormal basis $\{e_k\}_0^\infty$ and in the first place define the operator A for the unit vectors e_k by the equations

$$Ae_k = b_{k-1}e_{k-1} + a_k e_k + b_k e_{k+1} \quad (k = 0, 1, 2, \dots; \ b_{-1} = 0). \qquad \dots[4.1]$$

† See for instance the book by AHKIEZER and GLAZMAN[1].

Due to its linearity the operator A is then also determined for all finite vectors

$$g = \sum x_k e_k.$$

Since by virtue of the definition [4.1]

$$(Ae_m, e_n) = (e_m, Ae_n) \quad (m, n = 0, 1, 2, \ldots),$$

it follows that for any two finite vectors f and g we have the equality

$$(Af, g) = (f, Ag).$$

But the set of all finite vectors is dense in H. Therefore the operator A is symmetric.† Consequently the operator permits a closure. This closure is a minimal closed linear operator that satisfies condition [4.1], i.e. any other closed symmetric operator satisfying condition [4.1] must be an extension of this one. We shall consider precisely this minimal closed operator and there is no reason why we should not keep the notation A for it.

Further we introduce the adjoint operator A^*. According to the general rule the vector

$$g = \sum_0^\infty x_k e_k \qquad \ldots[4.2]$$

belongs to D_{A^*} if and only if the vector

$$g^* = \sum_0^\infty y_k e_k,$$

exists for which

$$(Ae_k, g) = (e_k, g^*) \quad (k = 0, 1, 2, \ldots).$$

With the aid of equations [4.1] these relations can be written as

$$(b_{k-1}e_{k-1} + a_k e_k + b_k e_{k+1}, g) = (e_k, g^*) \quad (k = 0, 1, 2, \ldots),$$

whence it follows that

$$y_k = b_{k-1}x_{k-1} + a_k x_k + b_k x_{k+1} \quad (k = 0, 1, 2, \ldots),$$

and therefore the vector [4.2] belongs to D_{A^*} if and only if

$$\sum_0^\infty | b_{k-1}x_{k-1} + a_k x_k + b_k x_{k+1} |^2 < \infty.$$

† A linear operator A in Hilbert space H is called symmetric if its domain of definition D_A is dense in H and if for any $f, g \in D$ one has the equality
$$(Af, g) = (f, Ag).$$

1.2. The next step consists in defining the deficiency index of the operator A. To do this we must seek non-trivial solutions of the equation

$$A^*g - \lambda g = 0 \qquad \qquad ...[4.3]$$

for Im $\lambda \neq 0$. Let a solution of this equation be the vector

$$g = \sum_0^\infty x_k e_k.$$

Then

$$\sum_0^\infty x_k(b_{k-1}e_{k-1} + a_k e_k + b_k e_{k+1}) = \lambda \sum_0^\infty x_k e_k,$$

whence it follows that

$$\lambda x_k = b_{k-1}x_{k-1} + a_k x_k + b_k x_{k+1} \quad (k = 1, 2, 3, ...),$$
$$\lambda x_0 = a_0 x_0 + b_0 x_1.$$

We have been led to the finite difference equation and boundary condition which were the subject of our discussion as early as Chapter 1 and which determine the polynomials of the first kind or the orthogonal polynomials associated with a given $\mathcal{J}$-matrix.

Thus

$$x_k = cP_k(\lambda) \quad (k = 0, 1, 2, ...),$$

where c is a constant. Hence one sees that equation [4.3] has a non-trivial solution if and only if

$$\sum_0^\infty |P_k(\lambda)|^2 < \infty,$$

and if this condition is satisfied, then the general solution of equation [4.3] is the vector

$$g = c \sum_0^\infty P_k(\lambda)e_k.$$

The linear manifold of solutions of equation [4.3] is denoted by $\mathfrak{N}_\lambda$. As we can see, its dimension (dim $\mathfrak{N}_\lambda$) is 0 or 1 according to whether the series

$$\sum_0^\infty |P_k(\lambda)|^2.$$

diverges or converges. It is known from elementary spectral theory of operators that dim $\mathfrak{N}_z$ maintains one and the same value throughout the half-plane Im $z > 0$ and also throughout the half-plane Im $z < 0$. The invariance theorem for $\mathscr{J}$-matrices proved in Section 3 of Chapter 1 can be considered as a consequence of this general statement of the invariance of dim $\mathfrak{N}_z$. In our case, since the polynomials $P_k(z)$ are real one has the equation dim $\mathfrak{N}_z = $ dim $\mathfrak{N}_{\bar{z}}$.

It follows from what has been said that the deficiency index of the operator A is $(0,0)$ or $(1,1)$. The first case occurs when the $\mathscr{J}$-matrix is of type D, the second when it is of type C.

1.3. By virtue of the equations [4.1] all integer non-negative powers of the operator A are determined at each of the unit vectors e_k, and all polynomials in the operator A are therefore also determined. Further we see that finding a unit vector e_k, starting from the unit vector e_1, by means of equations [4.1] requires the performance of the same algebraic operations as obtaining the polynomial $P_k(\lambda)$ from the polynomial $1 = P_0(\lambda)$. Therefore

$$e_k = P_k(A)e_0 \quad (k = 0, 1, 2, ...)$$

and to the expansion

$$\lambda^m = \sum_{j=0}^m \xi_j^{(m)} P_j(\lambda) \qquad ...[4.4]$$

there corresponds the representation

$$A^m e_0 = \sum_{j=0}^m \xi_j^{(m)} P_j(A)e_0 = \sum_{j=0}^m \xi_j^{(m)} e_j. \qquad ...[4.5]$$

On the basis of [4.4] we conclude that

$$s_{m+n} = S\{\lambda^{m+n}\} = S\{\lambda^m \lambda^n\} = \sum_{j=0}^m \sum_{k=0}^n \xi_j^{(m)} \xi_k^{(n)} \delta_{jk},$$

and it follows from [4.5] that

$$(A^m e_0, A^n e_0) = \sum_{k=0}^m \sum_{k=0}^n \xi_j^{(m)} \xi_k^{(n)} \delta_{jk}.$$

Therefore

$$(A^m e_0, A^n e_0) = s_{m+n} \quad (m, n = 0, 1, 2, ...).$$

This is the formula with the aid of which the moment problem is included in the problem of the spectral resolution of the operator A.

For brevity of expression we shall say that our moment problem is *associated* with the operator A or *generated* by this operator.

THEOREM **4.1.3.** *Let F_u be some spectral function† of an operator A. In this case the formula*

$$\sigma(u) = (F_u e_0, e_0)$$

gives a solution of the moment problem associated with the operator A and any solution of that moment problem can be represented in such a form.

Proof. I. Let F_u be a spectral function of the operator A. We shall prove that for any integral $k \geqq 0$ we have

$$(A^k e_0, f) = \int_{-\infty}^{\infty} u^k \, d(F_u e_0, f), \qquad \qquad ...[4.6]$$

where f is an arbitrary finite vector, and then, putting $f = e_0$ we shall obtain the equations

$$\int_{-\infty}^{\infty} u^k \, d\sigma(u) = \int_{-\infty}^{\infty} u^k \, d(F_u e_0, e_0) = (A^k e_0, e_0) = s_k.$$

For $k = 0$ and $k = 1$ equation [4.6] is obvious. In order to use induction we assume that equation [4.6] has been proved for $k = n-1$. Then we have

$$(A^n e_0, f) = (A^{n-1} e_0, Af) = \int_{-\infty}^{\infty} u^{n-1} \, d(F_u e_0, Af)$$

$$= \int_{-\infty}^{\infty} u^{n-1} \, d([F_u - F_0] e_0, Af). \qquad ...[4.7]$$

† We recall the following definitions.

A. By a *resolution of unity* we mean any one-parameter family of operators F_u ($-\infty \leqq u \leqq \infty$) satisfying the following conditions:

1) For $u_2 > u_1$ the difference $F_{u_2} - F_{u_1}$ is a bounded positive operator,

2) In the sense of convergence in norm on every element we have

$$F_{u-0} = F_u \quad (-\infty < u < \infty), \quad F_{-\infty} = 0. \quad F_{\infty} = E.$$

A′. A resolution of unity is called *orthogonal* if in addition

$$F_u F_v = F_w \quad (w = \min \{u, v\});$$

An orthogonal resolution of unity is denoted by E_u.

B. By a *spectral function* of a symmetric operator A we mean any resolution of unity F_u satisfying the following condition: for any $h \in H$ and any finite interval $\Delta = (u_1, u_2)$ the vector $F(\Delta)h = [F_{u_2} - F_{u_1}]h$ belongs to D_{A^*} and

$$(A^* F(\Delta) h, h') = \int_{\Delta} u \, d(F_u h, h'),$$

where h' is an arbitrary element of H.

B′. A spectral function is called *orthogonal* if it is an orthogonal resolution of unity.

But by definition of a spectral function we have

$$([F_u - F_0]e_0, Af) = (A^*[F_u - F_0]e_0, f) = \int_0^u t\, d(F_t e_0, f).$$

Inserting this into [4.7] we find that

$$(A^n e_0, f) = \int_{-\infty}^{\infty} u^{n-1} u\, d(F_u e_0, f) = \int_{-\infty}^{\infty} u^n\, d(F_u e_0, f),$$

and this is just equation [4.6] for $k = n$.

II. We now assume that $\sigma(u)$ is some solution of the moment problem. We assume that this solution satisfies the normalization condition

$$\sigma(u - 0) = \sigma(u).$$

For any real u and fixed k we then introduce the numbers

$$\int_{-\infty}^u P_k(t) P_n(t)\, d\sigma(t) = c_n(u, k) = c_n \quad (n = 0, 1, 2, \ldots),$$

which are the Fourier coefficients of the function

$$f(t) = \begin{cases} P_k(t) & (t \le u), \\ 0 & (t > u), \end{cases}$$

and therefore satisfy the inequality

$$\sum_0^\infty c_n^2 \le \int_{-\infty}^u [P_k(t)]^2\, d\sigma(t).$$

Therefore we can define the operator F_u for any real u at all e_k by putting

$$F_u e_k = \sum_{n=0}^\infty c_n(u, k) e_n.$$

The operator F_u can then be defined by linearity for all finite vectors. If

$$f = \sum_0^m x_k e_k,$$

then

$$(F_u f, f) = \int_{-\infty}^u \left| \sum_0^m x_k P_k(t) \right|^2 d\sigma(t),$$

whence one sees that

$$0 \le (F_u f, f) \le \int_{-\infty}^\infty \left| \sum_0^m x_k P_k(t) \right|^2 d\sigma(t) = \sum_0^m |x_k|^2 = (f, f).$$

This inequality shows that the operator F_u can be extended by continuity to the whole space H, maintaining its positive character, and also that the norm of F_u does not exceed 1. Also we note that for $u_1 < u_2$ the operator $F_{u_2} - F_{u_1}$ is positive. The properties

$$F_{u-0} = F_u, \quad F_{-\infty} = 0, \quad F_\infty = E$$

are also easy to verify. Thus F_u is a resolution of unity.

We note further that for any finite $u_1 < u_2$ we have

$$([F_{u_2} - F_{u_1}]e_k, Ae_n) = ([F_{u_2} - F_{u_1}]e_k, b_{n-1}e_{n-1} + a_ne_n + b_ne_{n+1})$$

$$= \int_{u_1}^{u_2} P_k(u)[b_{n-1}P_{n-1}(u) + a_nP_n(u) + b_nP_{n+1}(u)] \, d\sigma(u)$$

$$= \int_{u_1}^{u_2} uP_k(u)P_n(u) \, d\sigma(u) = \int_{u_1}^{u_2} u \, d(F_ue_k, e_n). \qquad ...[4.8]$$

Since this is true for any n we have

$$(F_{u_2} - F_{u_1})e_k \in D_{A^*}$$

and equation [4.8] can be written in the form

$$(A^*[F_{u_2} - F_{u_1}]e_k, e_n) = \int_{u_1}^{u_2} u \, d(F_ue_k, e_n).$$

Hence it follows directly that F_u is a spectral function of the operator A.

1.4. Of particular interest are the orthogonal spectral functions E_u of an operator A, which always exist, because the deficiency numbers of an operator A are equal. The problem arises as to what solutions of the moment problem are generated by these spectral functions.

THEOREM **4.1.4.** *A function $\sigma(u)$ is called an N-extremal solution of the moment problem if and only if*

$$\sigma(u) = (E_ue_0, e_0),$$

where E_u is an orthogonal spectral function of the operator A.

Proof. I. Let

$$\sigma(u) = (E_ue_0, e_0),$$

where E_u is an orthogonal spectral function of the operator A. In this case, if $\alpha \leq \beta$ we have the identity:

$$(E_\alpha e_0, e_0) = (E_\beta E_\alpha e_0, e_0) = (E_\alpha e_0, E_\beta e_0) = \sum_{k=0}^{\infty} (E_\alpha e_0, e_k)(e_k, E_\beta e_0).$$

Since

$$e_k = P_k(A)e_0 \quad (k = 0, 1, 2, \ldots),$$

we have

$$(E_u e_0, e_k) = (E_u e_0, P_k(A)e_0) = \int_{-\infty}^{\infty} P_k(t) \, d_t(E_t E_u e_0, e_0)$$

$$= \int_{-\infty}^{u} P_k(t) \, d(E_t e_0, e_0) = \int_{-\infty}^{u} P_k(t) \, d\sigma(t).$$

Therefore the identity in question can be represented in the form

$$\int_{-\infty}^{\alpha} d\sigma(u) = \sum_{k=0}^{\infty} \int_{-\infty}^{\alpha} P_k(u) \, d\sigma(u) \int_{-\infty}^{\beta} P_k(t) \, d\sigma(t). \quad \ldots[4.9]$$

We introduce the function

$$\theta_\mu = \theta_\mu(u) = \begin{cases} 1 & (-\infty < u \le \mu), \\ 0 & (\mu < u < \infty). \end{cases}$$

Then the relation [4.9] can be thought of as the Parseval relation for the pair of functions θ_α, θ_β. Since it is correct for any β and $\alpha \le \beta$ this equation is true for all step functions (sectionally constant functions). It is therefore true for any functions of L_σ^2. Therefore the set of polynomials is dense in L_σ^2 and therefore the solution $\sigma(u)$ is N-extremal.

II. We assume now that $\sigma(u)$ is an N-extremal solution of the moment problem. We shall use Theorem **4.1.3**, by virtue of which the operator A has a spectral function F_u such that

$$(F_u e_k, e_n) = \int_{-\infty}^{u} P_k(t) P_n(t) \, d\sigma(t) \quad \ldots[4.10]$$

for any k and n. Now we must only prove that for any real α and β we have

$$F_\beta F_\alpha = F_\gamma \quad (\gamma = \min \{\alpha, \beta\}).$$

For this it is sufficient to show that

$$(F_\beta F_\alpha e_k, e_n) = (F_\gamma e_k, e_n) \quad \ldots[4.11]$$

for any k and n.

But since the solution $\sigma(u)$ is N-extremal for every pair of functions from L_σ^2 one can state a Parseval inequality. We shall write it down for the functions

$$\theta_\alpha(u) P_k(u), \quad \theta_\beta(u) P_n(u).$$

We then find that

$$\int_{-\infty}^{\gamma} P_k(u)P_n(u)\,d\sigma(u)$$

$$= \sum_{j=0}^{\infty} \int_{-\infty}^{\alpha} P_k(u)P_j(u)\,d\sigma(u) \int_{-\infty}^{\beta} P_n(t)P_j(t)\,d\sigma(t). \quad ...[4.12]$$

Taking into account the general formula [4.10] we rewrite equation [4.12] in the form

$$(F_\gamma e_k, e_n) = \sum_{j=0}^{\infty} (F_\alpha e_k, e_j)(e_j, F_\beta e_n).$$

But evidently the right hand side is given by

$$(F_\alpha e_k, F_\beta e_n) = (F_\beta F_\alpha e_k, e_n)$$

and therefore the relation [4.11] is proved.

2. Symmetric operators which can be represented with the aid of $\mathscr{J}$-matrices

2.1. In the preceding section we have shown that any $\mathscr{J}$-matrix generates a certain symmetric operator. When we say " generates " we mean " generates as a minimal closed operator ". We can say of the operator that it can be represented with the aid of a $\mathscr{J}$-matrix.

Now it is natural to pose the question: how wide is the class of symmetric operators which can be obtained in this way? Since symmetric operators which can be generated by $\mathscr{J}$-matrices have a deficiency index $(0, 0)$ or $(1, 1)$ it is convenient to split up our question into the following two:

α) What are the conditions that must be satisfied by a self-adjoint operator A in order that a $\mathscr{J}$-matrix (evidently of type D), should exist, which generates the operator?

β) What are the conditions that must be satisfied by a closed symmetric operator A with a deficiency index $(1, 1)$ in order that a $\mathscr{J}$-matrix (of type C) should exist, which generates the operator?

Question α) was posed and solved by STONE[1]. His solution is rather simple and we can give it here. As regards question β), it can be solved in terms of the spectrum of any self-adjoint extension of the operator A. The final criterion is rather complicated, we shall give it in the Addenda to this chapter. We note that the statement and investigation of question β) is due to HAMBURGER[4].

2.2 In solving question α) the concept of a *simple spectrum* plays a part.

We give a definition concerning this: a self-adjoint operator A in a Hilbert space H has a *simple spectrum* if there exists such an element $g \in H$ (*generating element*) that the linear hull of the set of vectors $(E_\beta - E_\alpha)g$, is dense in H, where E_u is an orthogonal spectral function of the operator A, and (α, β) ranges over all intervals on the number axis.

In order to make this definition clear, we recall the corresponding situation in the finite dimensional case. Let A be a linear symmetric operator in n-dimensional space. Let $g_1, g_2, ..., g_n$ be the complete system of its characteristic vectors and $\lambda_1, \lambda_2, ..., \lambda_n$ the corresponding characteristic values. The numbers $\lambda_1, \lambda_2, ..., \lambda_n$ form the spectrum of the operator and this spectrum is called *simple* if all the λ_k are different. Any vector g of such an n-dimensional space can be represented in the form

$$g = \alpha_1 g_1 + \alpha_2 g_2 + \ ... \ + \alpha_n g_n. \qquad ...[4.13]$$

We choose a vector g in such a way that all α_k are non-zero. With this choice all the characteristic vectors g_k enter as components into the composition of the vector g. If we apply the operator A^m to the vector g, or else any linear function of the operator A we arrive at the vector

$$A^m g = \lambda_1^m \alpha_1 g_1 + \lambda_2^m \alpha_2 g_2 + \ ... \ + \lambda_n^m \alpha_n g_n, \qquad ...[4.14]$$

or in the general case to the vector

$$\phi(A)g = \phi(\lambda_1)\alpha_1 g_1 + \phi(\lambda_2)\alpha_2 g_2 + \ ... \ + \phi(\lambda_n)\alpha_n g_n. \qquad ...[4.14a]$$

In the case where the spectrum is not simple, for instance if $\lambda_1 = \lambda_2$ the vectors g_1 and g_2 enter into the composition of any one of the vectors [4.14] or [4.14a] in the same combination, $\alpha_1 g_1 + \alpha_2 g_2$, as in the initial vector [4.13]. Therefore, no matter how many vectors of type [4.14] or [4.14a] we choose, there will not be among them a full system of linearly independent vectors in the n-dimensional space in question. Conversely in the case of a simple spectrum it is very easy to choose functions $\phi_m(\lambda)$ $(m = 0, 1, ..., n-1)$ such that the vectors $\phi_m(A)g$ $(m = 0, 1, ... n-1)$ are linearly independent. For instance one can choose $\phi_m(A) = A^m$. The property we have here described is just what is taken as the basis of the generalization of the concept of a simple spectrum for self-adjoint operators in H. Here one takes as functions $\phi(A)$ the operators $E_\beta - E_\alpha$ where (α, β) ranges over all intervals on the number axis.

We now turn to some self-adjoint operator A which has a simple spectrum. We take any generating element g of this operator and then introduce the non-decreasing function $\omega(u) = (E_u g, g)$ $(-\infty \le u \le \infty)$

and construct the space L_ω^2. For any function $f(u) \in L_\omega^2$ the following operator integral has a meaning:

$$\int_{-\infty}^{\infty} f(u) \, dE_u g.$$

This integral is an element of the space H which we shall denote by f and which, by the definition of a function of a self-adjoint operator is the result of applying the operator $f(A)$ to the element g. Thus

$$f = f(A)g = \int_{-\infty}^{\infty} f(u) \, dE_u g.$$

With the aid of this formula we obtain a certain mapping of the whole space L_ω^2 on the whole space H. (We should remember that even the set of all sectionally constant functions $f(u)$ goes over into a set dense in H, since the spectrum of the operator A is simple and g is a generating element.) This mapping of the space L_ω^2 is an isometric mapping since

$$\| f \|^2 = (f,f) = \int_{-\infty}^{\infty} |f(u)|^2 \, d(E_u g, g) = \int_{-\infty}^{\infty} |f(u)|^2 \, d\omega(u).$$

We assume now that not only $f(u) \in L_\omega^2$ but also $uf(u) \in L_\omega^2$. In this case we have the pair of equations

$$f = f(A)g = \int_{-\infty}^{\infty} f(u) \, dE_u g, \quad Af(A)g = \int_{-\infty}^{\infty} uf(u) \, dE_u g.$$

Hence we see that in the isometric mapping of L_ω^2 on to H there corresponds to the operator A in H the operator of multiplying by the independent variable in L_ω^2.

We now show that if the self-adjoint operator A is generated by a $\mathscr{J}$-matrix of type D, then its spectrum is simple. Thus the simplicity of the spectrum of a self-adjoint operator is a necessary condition for it to be capable of representation with the aid of matrices of type D. It turns out, and this is the content of Stone's theorem which will be proved in the following section, that this condition is also sufficient.

Thus let the operator A be generated by a $\mathscr{J}$-matrix of type D. We take an element e_0 and construct all vectors $A^k e_0$ ($k = 0, 1, 2, \ldots$); the linear hull of these is dense in H, since it coincides with the linear hull of the unit vectors e_k.

For any vector $f \in$ H one can write down the equations

$$(A^k e_0, f) = \int_{-\infty}^{\infty} u^k \, d(E_u e_0, f) \quad (k = 0, 1, 2, \ldots).$$

It follows from what has been said that not all these integrals can be zero if $f \neq 0$. Therefore an element $f \neq 0$ does not exist for which $(E_u e_0, f) = 0$ for every u. But this just means that the linear hull of the set of all vectors of the form $(E_\beta - E_\alpha) e_0$ is dense in H, i.e. that the spectrum of the operator A is simple and that e_0 is its generating element.

2.3. THEOREM 4.2.3 (STONE[1]). *Any self-adjoint operator A with simple spectrum is generated by some $\mathscr{J}$-matrix of type D.*

Proof. Let A be a self-adjoint operator with simple spectrum, E_u an orthogonal resolution of unity belonging to it and h any generating vector. We put

$$g = \int_{-\infty}^{\infty} e^{-u^2} \, dE_u h.$$

All natural powers of the operator A have a meaning applied to the vector g:

$$A^k g = \int_{-\infty}^{\infty} u^k e^{-u^2} \, dE_u h.$$

We now prove that a vector $f \neq 0$ orthogonal to all the vector $A^k g$ ($k = 0, 1, 2, \ldots$) does not exist. Indeed by Section **2.2** any vector $f \in H$ may be represented in the form

$$f = \int_{-\infty}^{\infty} f(u) \, dE_u h, \qquad \qquad \ldots [4.15]$$

where $f(u) \in L^2_\omega$, $\omega(u) = (E_u h, h)$. But

$$(A^k g, f) = \int_{-\infty}^{\infty} u^k e^{-u^2} \, d(E_u h, f)$$

and by virtue of [4.15]

$$(f, E_t h) = \int_{-\infty}^{\infty} f(u) \, d_u(E_u h, E_t h) = \int_{-\infty}^{t} f(u) d(E_u, h) = \int_{-\infty}^{t} f(u) \, d\omega(u).$$

Therefore

$$(A^k g, f) = \int_{-\infty}^{\infty} u^k e^{-u^2} \overline{f(u)} \, d\omega(u) \quad (k = 0, 1, 2, \ldots).$$

If all these quantities were zero we would have the equations

$$\int_{-\infty}^{\infty} u^k e^{-u^2} \, d\Omega(u) = 0 \quad (k = 0, 1, 2, \ldots),$$

where

$$\Omega(u) = \int_{-\infty}^{u} f(t)\, d\omega(t) - C,$$

and C is an arbitrary constant. Integrating by parts we obtain

$$\int_{-\infty}^{\infty} \Omega(u)(ku^{k-1} - 2u^{k+1})e^{-u^2}\, du = 0 \quad (k = 0, 1, 2, \ldots). \qquad \ldots[4.16]$$

If the constant C is chosen so that

$$\int_{-\infty}^{\infty} \Omega(u)e^{-u^2}\, du = 0,$$

it follows from $[4.16]$ that

$$\int_{-\infty}^{\infty} \Omega(u)u^k e^{-u^2}\, du = 0 \quad (k = 0, 1, 2, \ldots). \qquad \ldots[4.17]$$

By virtue of the well known completeness theorem for Hermite polynomials, equations $[4.17]$ lead to the conclusion that $\Omega(u) = 0$ and we obtain the contradiction

$$0 \neq \int_{-\infty}^{\infty} |f(u)|^2\, d\omega(u) = \int_{-\infty}^{\infty} \overline{f(u)}\, d\Omega(u) = 0.$$

Thus it is proved that a vector $f \neq 0$, orthogonal to all the $A^k g$ $(k = 0, 1, 2, \ldots)$ does not exist. Hence it follows that the set of all vectors of the form $(E_\beta - E_\alpha)g$ is dense in H and therefore that the vector g as well as h is a generating vector for the operator A. Replacement of g by αg $(\alpha \neq 0)$ does not change anything. Therefore we assume that $\| g \| = 1$.

We now orthogonalize the sequence of vectors $\{A^k g\}_0^\infty$. We then obtain a certain orthonormal base $\{e_k\}_0^\infty$ in H, all elements of which belong to D_A. Here

$$e_k = P_k(A)e_0 = \int_{-\infty}^{\infty} P_k(u)\, dE_u e_0, \quad e_0 = g,$$

and the sequence of polynomials $\{P_k(u)\}_0^\infty$ forms an orthonormal base in L_σ^2 where $\sigma(u) = (E_u e_0, e_0)$. Since

$$(Ae_k, e_i) = \int_{-\infty}^{\infty} u P_k(u) P_i(u)\, d\sigma(u),$$

the matrix $\| (Ae_k, e_i) \|_{i,\,k=0}^\infty$ of the operator A referred to the base $\{e_k\}_0^\infty$ is a $\mathcal{J}$-matrix.

We now show that the minimal closed operator B generated by this matrix is a self-adjoint operator. Hence it will follow that $B = A$, since otherwise the operator A would be an extension of the operator B, which is impossible. This will conclude the proof of the theorem.

In order to prove that the operator B is self-adjoint one must verify that the range of values of each of the two operators $B \pm iE$ coincides with the whole space H. Putting it differently, it must be proved that a vector ϕ orthogonal to all vectors $(B - iE)e_k$ $(k = 0, 1, 2, ...)$, is necessarily zero and that the same is true for any vector ψ orthogonal to all vectors $(B + iE)e_k$.

We assume that the vector ϕ is orthogonal to all vectors $(B - iE)e_k$. On the other hand this vector must be representable in the form

$$\phi = \int_{-\infty}^{\infty} \phi(u)\, dE_u h, \qquad \phi(u) \in L_\omega^2.$$

Therefore

$$0 = ((B - iE)e_k, \phi) = ((A - iE)e_k, \phi) = ((A - iE)P_k(A)e_0, \phi)$$

$$= \int_{-\infty}^{\infty} (u - i)P_k(u)\overline{\phi(u)}e^{-u^2}\, d\omega(u). \qquad ...[4.18]$$

Now we put

$$\Omega(u) = \int_{-\infty}^{u} (t + i)\phi(t)e^{-\frac{1}{2}t^2}\, d\omega(t) - C,$$

where C is a constant. In this case equations $[4.18]$ take on the form

$$\int_{-\infty}^{\infty} P_k(u)e^{-\frac{1}{2}u^2}\, d\Omega(u) = 0 \quad (k = 0, 1, 2, ...),$$

or

$$\int_{-\infty}^{\infty} u^k e^{-\frac{1}{2}u^2}\, d\Omega(u) = 0 \quad (k = 0, 1, 2, ...).$$

Hence, repeating the argument already used once, we find that with a suitable choice of the constant C we have $\Omega(u) = 0$. Therefore

$$\int_{-\infty}^{\infty} |u - i|^2 e^{-\frac{1}{2}u^2} |\phi(u)|^2\, d\omega(u) = \int_{-\infty}^{\infty} (u + i)\phi(u)\, d\overline{\Omega(u)} = 0,$$

whence

$$\int_{-\infty}^{\infty} |\phi(u)|^2\, d\omega(u) = 0,$$

i.e. the vector ϕ is zero. Analogously one proves that the vector ψ is also zero.

The theorem is therefore completely proved.

2.4. To conclude this section we establish a certain necessary (but by no means sufficient) condition for a closed symmetric operator to be representable with the aid of a $\mathcal{J}$-matrix of type C. This condition is that the operator must be simple.

We digress a little to recollect the definition of a simple symmetric operator.

Let T be any closed linear operator in H. If the space $H_1 \subset H$ and its orthogonal complement $H \ominus H_1 = H_2$ are invariant subspaces of T and if projection into H_1 does not take an element f out of D_T, one says that the space H_1 reduces the operator T. The justification of this name is the following fact:

If the space H_1 reduces T then given any $f \in D_T$ we have

$$Tf = T_1 f_1 + T_2 f_2,$$

where f_1 and f_2 are the projections of f into H_1 and H_2, and T_1 and T_2 are the parts of T lying in H_1 and H_2.

A symmetric operator A is called *simple* if a space reducing it, in which a part of this operator is a self-adjoint operator, does not exist.

THEOREM **4.2.4.** *If a closed symmetric operator A is represented by a $\mathcal{J}$-matrix of type C, then A is a simple operator.*

Proof. The deficiency index of the operator A is $(1, 1)$, and any self-adjoint extension of this operator generates an N-extremal solution of the moment problem which has only discrete points of increase. These discrete points form the spectrum of this particular extension of the operator A. The spectra of two different extensions of this kind have no points in common.

Indeed each such spectrum is formed by the zeros of the entire function

$$B(z)t - D(z),$$

where t is a real constant and two functions of this form with different t have no common zeros.

It follows that the operator A cannot be reducible since if it were it would follow that all self-adjoint extensions of the operator would have at least one common characteristic vector, and with it a common characteristic value.

Thus the theorem is proved.

It can be proved (see for instance AHKIEZER and GLAZMAN[1], sections 69 and 81) that if a simple symmetric operator has deficiency index (1, 1) then any self-adjoint extension of this operator has a simple spectrum.

Using Theorem 4.2.3 it follows that any self-adjoint extension of a simple closed symmetric operator A with deficiency index (1, 1) can be represented by a $\mathscr{J}$-matrix, and in fact one of type D. However this does not mean by any means that the operator A itself can be represented by a $\mathscr{J}$-matrix. In other words, our statement is not an inversion of Theorem 4.2.4 and does not give an answer, in combination with that theorem, to the question β) stated at the beginning of this section.

3. Integral representations of a positive functional

3.1. In this section we shall again examine the problem of moments (both the power problem and the generalized one) as a problem of continuing a positive functional. However in contrast to Section 6 of Chapter 2 where general constructs of functional analysis were used we shall now use special techniques based on the spectral theory of operators. If one has a power moment problem in mind then the set of all polynomials must be taken as the domain of definition of the given functional. However, we shall not consider ourselves bound by this limitation and shall discuss the problem stated in a general form. The first discussion of this kind is due to LIVSHITS[1]. Somewhat later KREIN[6, 8] developed Livshits' results somewhat further and gave them greater generality. This was the origin of Krein's *method of directing functionals* which proved extremely effective for certain problems in the spectral theory of differential operators.

We begin with a precise formulation of our problem. For this we must begin by describing those linear manifolds of functions over which we shall form functionals and also define precisely how we shall give meaning to the terms positive (non-negative) applied to a functional.

We take as our starting point a certain linear aggregate G of complex valued continuous functions $f(s)$ $(-\infty < s < \infty)$. We consider all possible products $f(s)\overline{g(s)}$, where $f(s), g(s) \in$ G, and we denote by $\mathfrak{M}$ the linear hull of all such products. This linear hull will just be the domain of definition of the functionals. A functional $\mathfrak{P}$, linear (in the algebraic sense), given on $\mathfrak{M}$ will be called *positive* (and correspondingly non-negative) if for any element $\phi \in \mathfrak{M}$ of the form $\phi = f(s)\overline{f(s)}$, where $f(s) \in$ G and $f(s) \not\equiv 0$, the inequality $\mathfrak{P}(\phi) > 0$ (or $\mathfrak{P}(\phi) \geqq 0$) holds.

Thus this definition of the positive character of a functional involves not all functions $\phi = \phi(s)$ belonging to $\mathfrak{M}$ for which $\phi(s) \geqq 0$ $(-\infty < s < \infty)$, but only all squares. This distinction is sometimes only apparent, for instance in the case when G is the set of all polynomials.

The formulation of the general moment problem is then the following: *find the conditions under which a given non-negative functional* P *defined over* $\mathfrak{M}$ *permits the integral representation*

$$P\{\phi\} = \int_{-\infty}^{\infty} \phi(t)\,d\sigma(t) \quad (\phi \in M) \qquad \ldots[4.19]$$

with a non-decreasing function $\sigma(t)$ $(-\infty < t < \infty)$. Evidently it is sufficient to obtain the representation [*4.19*] for functions $\phi(s) = f(s)\overline{f(s)}$, where $f(s) \in G$. We shall call this problem the generalized Livshits moment problem. It becomes the ordinary power moment problem if G is the aggregate of all polynomials in s.

In the following we shall discuss only sufficient conditions for the Livshits moment problem to be soluble. It is worth noting that if this problem is soluble, then the inequality $P\{\phi\} \geqq 0$ must hold for any real non-negative function $\phi(s) \in \mathfrak{M}$.

To include the generalized moment problem in the spectral theory of operators we must have a Hilbert space. In contrast to Section 1 of this chapter we cannot here use the usual (abstract) notion of a Hilbert space, but must construct the space with the aid of the data of our problem. To do this we perform a metrization of the linear aggregate G by means of the positive functional P, namely by defining for any two elements $f, g \in G$ the scalar product according to the formula

$$(f, g) = P\{f(s)\overline{g(s)}\}.$$

This scalar product evidently satisfies the following conditions:

$$1° \qquad \overline{(f, g)} = (g, f),$$
$$2° \qquad (f_1 + f_2, g) = (f_1, g) + (f_2, g),$$
$$3° \qquad (\alpha f, g) = \alpha(f, g),$$
$$4° \qquad (f, f) \geqq 0.$$

If the functional P is positive then the equality sign in property 4 is possible only if $f(s) \equiv 0$. However, we shall assume only that the functional P is non-negative. Therefore the existence of elements $f_0(s) \not\equiv 0, f_0(s) \in G$, for which $(f_0, f_0) = 0$ is not excluded. We show first that all elements f_0 of this kind form a linear aggregate $G_0 \subset G$. To do this we write down the Schwartz-Bunyakovskii inequality

$$| (f_0, g) | \leqq \sqrt{(f_0, f_0)} \sqrt{(g, g)}$$

which follows from property 4. It shows that the equation $(f_0, f_0) = 0$ is only possible for some element $f_0 \in G$ provided that for any $g \in G$

the equality $(f_0, g) = 0$ holds and this last relation is homogeneous and linear with respect to f_0. We now agree to consider as identical in G any two elements f_1 and f_2 for which $f_1 - f_2 \in G_0$. If we do this, the scalar product will satisfy all the requirements imposed on it, without any exceptions. Having done this, we close the linear manifold G. As a result we obtain a Hilbert space which we denote by H.

We impose one additional restriction on the set G, namely the requirement that the aggregate D of all those elements $g(s) \in G$ for which $sg(s) \in G$ should be dense in G, and therefore also in H. Under these conditions we can introduce in H the operator of multiplication by the independent variable s, having defined it initially on the aggregate D. This operator is symmetric since the set D is dense in H and

$$(f(s), sg(s)) = \mathcal{P}\{f(s)\overline{sg(s)}\} = (sf(s), g(s))$$

for any $f(s), g(s) \in D$. Therefore we can close it. This closure is just what we shall mean by the operator A. To verify the correctness of this definition we must simply show that the result of applying the operator A will be the same for all elements which by virtue of our definition are assumed to be identical. In other words we must verify that if $(f_0(s), f_0(s)) = 0$, then $(sf_0(s), sf_0(s)) = 0$. Evidently it is sufficient to verify this fact for $f_0(s) \in D$. But the equation $(f_0(s), f_0(s)) = 0$ means that $(f_0(s), g(s)) = 0$ for any $g(s) \in G$ and in particular for any $g(s) = sh(s)$, $h(s) \in D$. Thus for any $h(s) \in D$ we have

$$(sf_0(s), h(s)) = (f_0(s), sh(s)) = 0.$$

Hence, by virtue of the fact that D is dense in G the required equation $(sf_0(s), sf_0(s)) = 0$ follows.

Now we assume temporarily that the moment problem in question is solved and therefore that for any $f(s), g(s) \in G$ we have

$$(f(s), g(s)) = \int_{-\infty}^{\infty} f(t)\overline{g(t)} \, d\sigma(t).$$

We see that the functions f and g appear here in two roles: on the one hand as elements of the space H (in which case we denote them by $f(s), g(s)$) on the other hand as elements of the space L_σ^2 (in which case they are denoted by $f(t), g(t)$). If the manifold G contains the function $e = e(s)$ which is identically equal to 1, then $f(s)$ and $g(s)$ can be considered as $f(A)e$ and $g(A)e$. In the case of the power moment problem the functions $f(s)$ and $g(s)$ are polynomials and therefore one can write down the equation

$$f(s) - f(t)e = [f(A) - f(t)E]e = (A - tE)g_t(A)e = (A - tE)g_t(s),$$

where $g_t(s)$ is a polynomial in s (i.e. an element of the Hilbert space), depending on the parameter t. This equation was seen by Livshits to be the key to the generalized moment problem.

3.2. The result that we shall need in order to generalize the moment problem is a particular case of a more general proposition which relates to any closed symmetric operator A in an arbitrary Hilbert space H. This proposition is:

THEOREM **4.3.2** (LIVSHITS[1]). *Let A be a closed symmetric operator in a Hilbert space* H *and F_t a spectral function of A. Assume further that to any two elements f, v one can adjoin a continuously differentiable function $f(t)(-\infty < t < \infty)$ such that for any $t \in (-\infty, \infty)$ an element $g_t \in D_A$, can be found for which*

$$f - f(t)v = (A - tE)g_t. \qquad \ldots[4.20]$$

In this case

$$f = \int_{-\infty}^{\infty} f(t)\, dF_t v; \quad (f, f) = \int_{-\infty}^{\infty} |f(t)|^2\, d(F_t v, v). \quad \ldots[4.21]$$

Proof. We begin by showing that this theorem reduces to a special case of itself which we propose to call a lemma and which we shall then prove.

LEMMA **4.3.2.** *Let A be a self-adjoint operator with the spectral function E_t. Assume further that to two elements f, $v \in$ H one can adjoin a continuously differentiable function $f(t)$ ($\alpha < t < \beta$) such that for any $t \in (\alpha, \beta)$ an element $g_t \in D_A$, may be found for which*

$$f - f(t)v = (A - tE)g_t. \qquad \ldots[4.20a]$$

In this case the representation

$$(E_{t''} - E_{t'}) = \int_{t'}^{t''} f(t)\, dE_t v. \qquad \ldots[4.21]$$

is valid in any interval $(t', t'') \subset (\alpha, \beta)$.

Let us assume for a moment that the lemma is proved. For the case that the operator A is self-adjoint Theorem **4.3.2** then follows directly. Indeed in this case we can use the lemma for $\alpha = -\infty$ and $\beta = \infty$ and this means that we can put $t' \to -\infty$, $t'' \to \infty$ in equation [4.21a]. But this gives us the first of the relations [4.21] to be proved and for the case that the spectral function E_t is orthogonal the second relation is a consequence of the first.

To prove Theorem **4.3.2** in the case of an arbitrary symmetric operator A we use the theorem due to NAIMARK[2] according to which any spectral function F_t of the operator A has corresponding to it a self-adjoint operator $\mathscr{A}$ in a certain extension $\mathscr{H}$ of the space H such that for $h \in D_A$

$$\mathscr{A}h = Ah$$

and

$$F_t = \mathscr{P}\mathscr{E}_t \quad (-\infty \leqq t \leqq \infty),$$

where the spectral function $\mathscr{E}_t$ belongs to the operator $\mathscr{A}$ and $\mathscr{P}$ is the projection operator from $\mathscr{H}$ on H.

By virtue of what has been proved we have

$$f = \int_{-\infty}^{\infty} f(t) \, d\mathscr{E}_t v; \quad (f, f) = \int_{-\infty}^{\infty} |f(t)|^2 \, d(\mathscr{E}_t v, v).$$

Applying the operator $\mathscr{P}$ to the first equation we find that

$$f = \mathscr{P}f = \int_{-\infty}^{\infty} f(t) \, d\mathscr{P}\mathscr{E}_t v = \int_{-\infty}^{\infty} f(t) \, dF_t v,$$

and on the other hand the second equation can be rewritten in the form

$$(f, f) = \int_{-\infty}^{\infty} |f(t)|^2 \, d(F_t v, v),$$

since

$$(\mathscr{E}_t v, v) = (\mathscr{E}_t v, \mathscr{P}v) = (\mathscr{P}\mathscr{E}_t v, v) = (F_t v, v).$$

Thus the theorem is proved if the lemma is correct.

Proof† of lemma **4.3.2.** We introduce the vector function

$$\omega_\tau = \int_\tau^{t''} dE_t f - \int_\tau^{t''} f(t) \, dE_t v \quad (t' \leqq \tau \leqq t''), \qquad \ldots [4.22]$$

which evidently becomes zero for $\tau = t''$. Therefore it is sufficient for the proof of the lemma to convince oneself that ω_τ is independent of τ. To do this we prove that for any $\tau \in (t', t'')$

α) ω_τ is continuous in norm

β) The equality holds

$$\lim_{\delta \to 0} \frac{1}{\delta} \| \omega_{\tau \pm \delta} - \omega_\tau \| = 0 \quad (\delta > 0). \qquad \ldots [4.23]$$

Indeed from this it would follow that ω_τ has a vanishing strong derivative at every point $\tau \in (t', t'')$ and this means that ω_τ is independent of τ.

† In Livfshits' short note the proof is only sketched and one statement is not justified. In connection with the proof given here see the article by KREIN[8].

From the definition $[4.22]$ and also from $[4.20a]$ we have

$$\omega_\tau - \omega_{\tau+\varepsilon} = \int_\tau^{\tau+\varepsilon} dE_t f - \int_\tau^{\tau+\varepsilon} f(t)\, dE_t v$$

$$= \int_\tau^{\tau+\varepsilon} d_t E_t (A - \tau E) g_\tau - \int_\tau^{\tau+\varepsilon} [f(t) - f(\tau)]\, dE_t v$$

$$= \int_\tau^{\tau+\varepsilon} (t - \tau)\, d_t E_t g_\tau - \int_\tau^{\tau+\varepsilon} [f(t) - f(\tau)]\, dE_t v.$$

Therefore

$$\| \omega_\tau - \omega_{\tau+\varepsilon} \| \leqq | \varepsilon | \cdot \| (E_{\tau+\varepsilon} - E_\tau) g_\tau \|$$

$$+ | \varepsilon | \cdot | f'(\tau + \theta\varepsilon) | \cdot \| (E_{\tau+\varepsilon} - E_\tau) v \| , \quad \dots [4.24]$$

where $0 < \theta < 1$. Statement α) follows directly from this result. Further taking $\varepsilon = -\delta < 0$, we write $[4.24]$ in the form

$$\frac{1}{\delta} \| \omega_{\tau-\delta} - \omega_\tau \| \leqq \| (E_\tau - E_{\tau-\delta}) g_\tau \| + | f'(\tau - \theta\delta) | \cdot \| (E_\tau - E_{\tau-\delta}) v \| ,$$

and in view of the normalization $E_{t-0} = E_t$ the second of relations $[4.23]$ follows. To prove the first of these relations we use the fact that the vector function ω_τ is continuous and write $[4.24]$ for $\varepsilon = \delta > 0$ in the form

$$\frac{1}{\delta} \| \omega_{\tau+\delta} - \omega_\tau \| \leqq \| (E_{\tau+\delta} - E_{\tau+0}) g_\tau \|$$

$$+ | f'(\tau + \theta\delta) | \cdot \| (E_{\tau+\delta} - E_{\tau+0}) v \| .$$

The right hand side of this inequality evidently tends to zero with δ.

3.3. We now turn to the generalized moment problem formulated in Section **3.1.**

THEOREM **4.3.3** (LIVFSHITS[1]). *The moment problem* $[4.19]$ *is certainly soluble if*

a) $1 \in G,$

b) *The set* D *of all functions* $g(s) \in G,$ *for which* $sg(s) \in G$ *is dense in* G *with the norm*

$$\| f \| = \sqrt{ \mathsf{P}\{f(s)\overline{f(s)}\} } \quad [f(s) \in G],$$

c) *From* $f(s) \in G$ *it follows that for any real* t *the function of* s

$$\frac{f(s) - f(t)}{s - t}$$

belongs to G.

Proof. We construct the Hilbert space H as indicated in Section 3.1, and in it we introduce the closed symmetric operator of multiplication by the independent variable. Then we apply Theorem **4.3.2**, taking as the element v the function $1 \in G$ and as the function $f(t)$ the value at the point t of the function $f(s) \in G$. Then the requirement [*4.20*] takes on the form

$$f(s) - f(t) \cdot 1 = (s-t)g_t(s)$$

and therefore it is satisfied, because both functions of s

$$g_t(s) = \frac{f(s)-f(t)}{s-t}, \quad sg_t(s) = tg_t(s) + f(s) - f(t) \cdot 1$$

belong to G.

Thus the theorem is proved.

It may be verified easily that the theorem just proved gives a criterion for the solubility of Hamburger's problem, and not only this. Another application of Theorem **4.3.3** will be discussed in the following chapter.

3.4. In this section we shall give a description of the method of directing functionals due to Krein, which was mentioned at the very beginning of the section.

We shall consider a Hilbert space H and in it a certain symmetric operator A. We assume that in this space a linear functional $\Phi(f; t)$, $f \in H$, is given, which depends on the real parameter $t \, (-\infty < t < \infty)$ and which satisfies the following conditions:

a) for a fixed element $f \in H$ the functional $\Phi(f; t)$ is a continuously differentiable function of t:

b) for any finite interval (α, β) of the real axis an element $v \in H$ can be found such that

$$\Phi(v; \, t) \neq 0 \quad (\alpha < t < \beta);$$

c) the necessary and sufficient condition for the equation

$$A\chi - t_0\chi = f_0$$

to be soluble is that the functional $\Phi(f; t)$ should become zero for the element $f_0 \in H$ and the particular real value t_0.

We note at once that in Krein's theory a more general case is considered when the criterion for the equation

$$A\chi - t\chi = f$$

to be soluble requires the introduction of a whole system of directing functionals; however we shall not deal with this generalization.

We assume that a finite interval (α, β) of the real axis has somehow been chosen and that v is that element of the space which according to condition b) corresponds to the interval (α, β). We take an arbitrary vector $f \in H$ and consider the difference

$$f_1 = f - \frac{\Phi(f; t)}{\Phi(v; t)} v.$$

Since $\Phi(f_1; t) = 0$ $(\alpha < t < \beta)$, then according to c) the equation

$$f - \frac{\Phi(f; t)}{\Phi(v; t)} v = (A - tE)g_t$$

is true for any t in the interval (α, β). We see that the ratio

$$\frac{\Phi(f; t)}{\Phi(v; t)}$$

plays the rôle of the function $f(t)$ occurring in Lemma **4.3.2**. In contrast to the situation in the Livfshits case, both the element v and the function $f(t)$ may now change if the interval (α, β) changes.

THEOREM **4.3.4** (KREIN[8]). *If A is a symmetric operator in H which possesses a directing functional $\Phi(f, t)$, then there exists at least one non-decreasing function $\sigma(t)$ $(-\infty < t < \infty)$ such that for any $f, g \in$ H*

$$(f, g) = \int_{-\infty}^{\infty} \Phi(f; t)\overline{\Phi(g; t)} \, d\sigma(t). \qquad \ldots [4.25]$$

Proof. We take the orthogonal spectral function E_u of the operator A if it is self-adjoint and of any self-adjoint extension of that operator (if necessary going beyond the Hilbert space H) otherwise. We then use Lemma **4.3.2**, which is possible in view of the remark made at the beginning of this section. By virtue of this lemma and of the definition of a directing functional we can associate with any finite interval (α, β) an element $v \in$ H such that for any interval $(t', t'') \in (\alpha, \beta)$ and any $h \in$ H the equality

$$(E_{t''} - E_{t'})h = \int_{t'}^{t''} \frac{\Phi(h; t)}{\Phi(v; t)} \, dE_t v$$

holds. Taking $h = f$ and $h = g$ and forming the scalar product we obtain:

$$([E_{t''} - E_{t'}]f, g) = \int_{t'}^{t''} \frac{\Phi(f; t)}{\Phi(v; t)} \frac{\overline{\Phi(g; t)}}{\overline{\Phi(v; t)}} \, d(E_t v, v)$$

$$= \int_{t'}^{t''} \Phi(f; t)\overline{\Phi(g; t)} \, d\sigma(t), \qquad \ldots [4.26]$$

where

$$\sigma(t) = \int_\gamma^t \frac{d(E_\tau v, v)}{|\Phi(v;\tau)|^2} = \int_\gamma^t \frac{d_\tau([E_\tau - E_\gamma]v, v)}{|\Phi(v;\tau)|^2}, \qquad \dots[4.27]$$

and where γ is some point in the interval (α, β).

To complete the proof we must convince ourselves that the non-decreasing function $\sigma(t)$ ($\alpha < t < \beta$) is independent of the choice of the element v and therefore in particular does not change within the interval (α, β) but is merely continued if one goes over from that interval to a greater interval containing it. Indeed if this is proved then one can go to the limit $t' \to -\infty$, $t'' \to \infty$ in equation $[4.26]$, so obtaining equation $[4.25]$.

Now, assume that in place of the element v one can take an element v^*. Then we shall have in addition to $[4.26]$ and $[4.27]$ the equations

$$([E_{t''} - E_{t'}]f, g) = \int_{t'}^{t''} \Phi(f;t)\overline{\Phi(g;t)}\, d\sigma^*(t), \qquad \dots[4.27a]$$

$$\sigma^*(t) = \int_\gamma^t \frac{d_\tau([E_\tau - E_\gamma]v^*, v^*)}{|\Phi(v^*;\tau)|^2}. \qquad \dots[4.27b]$$

We write down equation $[4.26]$ for $f = g = v^*$, $t'' = \tau$, $t' = \gamma$:

$$([E_\tau - E_\gamma]v^*, v^*) = \int_\gamma^\tau |\Phi(v^*;t)|^2\, d\sigma(t).$$

Inserting this into $[4.27a]$ we find that

$$\sigma^*(t) = \int_\gamma^t \frac{1}{|\Phi(v^*;\tau)|^2}\, d_\tau\left[\int_\gamma^\tau |\Phi(v^*;\xi)|^2\, d\sigma(\xi)\right] = \int_\gamma^t d\sigma(\tau),$$

whence it follows that $\sigma^*(t) = \sigma(t)$.

Thus the theorem is proved. An example of its application is given below.

Addenda and Problems†

1. DEFINITION 1. We shall say that the function $q(z)$ belongs to the class $\mathfrak{A}$ if it is entire and real, if it has an infinite set of zeros λ_j ($j = 1, 2, 3, \dots$) all of which are real and simple, if the absolutely convergent expansion

$$\frac{1}{q(z)} = \sum_1^\infty \frac{1}{q'(\lambda_j)(z - \lambda_j)} \qquad \dots[4.28]$$

† Examples 1–6 of this section are based on HAMBURGER's[4] results (see also HAMBURGER[5]) and on Krein's theorem (Addenda and Problems to Chapter 3 example 11).

is valid and if all the series

$$\sum_{j=1}^{\infty} \frac{\lambda_j^m}{q'(\lambda_j)} \quad (m = 0, 1, 2, \ldots). \qquad \ldots[4.29]$$

are convergent.

LEMMA 1. *If* $q(z) \in \mathfrak{A}$, *then the series* $[4.29]$ *converge absolutely, the inequality*

$$\sum_{j=1}^{\infty} \frac{\lambda_j^m}{q'(\lambda_j)} = 0 \quad (m = 0, 1, 2, \ldots) \qquad \ldots[4.29a]$$

holds and for any polynomial $S(z)$ *the following absolutely convergent expansion is valid*

$$\frac{S(z)}{q(z)} = \sum_{1}^{\infty} \frac{S(\lambda_j)}{q'(\lambda_j)(z - \lambda_j)}. \qquad \ldots[4.28a]$$

Proof. Krein's theorem tells us that the absolute convergence of the expansion $[2.48]$ implies that $q(z)$ is an entire function of exponential type. Therefore

$$\sum_{1}^{\infty} \frac{1}{1 + \lambda_j^2} < \infty.$$

Further, by virtue of the fact that the series $[4.29]$ converge, we have

$$\lim_{j \to \infty} \frac{\lambda_j^m (1 + \lambda_j^2)}{q'(\lambda_j)} = 0 \quad (m = 0, 1, 2, \ldots)$$

and therefore

$$\left| \frac{\lambda_j^m}{q'(\lambda_j)} \right| = \left| \frac{\lambda_i^m (1 + \lambda_j^2)}{q'(\lambda_j)} \right| \cdot \frac{1}{1 + \lambda_j^2} < \frac{C_m}{1 + \lambda_j^2} \quad (j = 1, 2, 3, \ldots),$$

whence one sees that the series $[4.29]$ converge absolutely.

The first statement is thus proved. Using it, one obtains from $[4.28]$ the asymptotic expansion

$$\frac{1}{q(z)} \sim \sum_{0}^{\infty} \frac{c_m}{z^{m+1}} \quad \left[c_m = \sum_{j=1}^{\infty} \frac{\lambda_j^m}{q'(\lambda_j)} \right],$$

which is valid for all angles within the limits

$$\delta \leqq \arg z \leqq \pi - \delta, \quad \pi + \delta \leqq \arg z \leqq 2\pi - \delta \quad \left(0 < \delta < \frac{\pi}{2} \right). \qquad \ldots[4.30]$$

Since $q(z)$ is an entire function of finite order and not a polynomial all the c_m are zero† i.e. equations $[4.29a]$ are true.

In order to prove the correctness of the absolutely convergent expansion $[4.28a]$ we multiply both sides of the series $[4.28]$ by $S(z)$. This gives the equation

$$\frac{S(z)}{q(z)} = \sum_1^\infty \frac{S(z) - S(\lambda_j)}{q'(\lambda_j)(z - \lambda_j)} + \sum_1^\infty \frac{S(\lambda_j)}{q'(\lambda_j)(z - \lambda_j)},$$

and it remains to note that by virtue of the equations $[4.28a]$ the first term of the right hand side vanishes.

LEMMA 2. *If $\{\lambda_j\}_1^\infty$ is the sequence of all zeros of the function $q(z)$ of class $\mathfrak{A}$, and the numbers $\mu_j > 0$ $(j = 1, 2, 3, ...)$ are such that*

$$\sum_{j=1}^\infty \mu_j \lambda_j^{2m} < \infty \quad (m = 0, 1, 2, ...)$$

and if

$$\sum_1^\infty \frac{1}{\mu_j (1 + \lambda_j^2)[q'(\lambda_j)]^2} < \infty, \qquad ...[4.31]$$

then the moments

$$s_m = \sum_{j=1}^\infty \mu_j \lambda_j^m \quad (m = 0, 1, 2, ...) \qquad ...[4.32]$$

generate an indeterminate Hamburger problem.

(The sequence $\{\mu_j\}_1^\infty$, may evidently be assumed normalized by means of the condition

$$\sum_1^\infty \mu_j = 1.$$

† This proof is rather simple. We assume that c_m is the first non-vanishing coefficient. We then have within the angles $[4.30]$

$$\lim_{z \to \infty} \frac{z^{n+1}}{q(z)} = c_n \neq 0.$$

In this case

$$q^*(z) = \frac{1}{z^{n+1}} \left\{ q(z) - q(0) - \frac{z}{1!} q'(0) - ... - \frac{z^n}{n!} q^{(n)}(0) \right\}$$

is again an entire transcendental function of finite order $\leqslant \rho$ (in our case $\rho = 1$) for which we have, within the ranges of angles $[4.30]$,

$$\lim_{z \to \infty} q^*(z) = \frac{1}{c_n}.$$

Therefore $q^*(z)$ is bounded within the angles $[4.30]$. Taking $\delta < \pi/2\rho$, we find from the Phragmén-Lindelöf theorem that the function $q^*(z)$ is bounded within the angles

$$-\delta \leqslant \arg z \leqslant \delta, \quad \pi - \delta \leqslant \arg z \leqslant \pi + \delta.$$

Therefore $q^*(z) = $const, which contradicts the fact that it is transcendental.

In the following we shall assume that this condition is satisfied.)

Proof. We introduce the sectionally constant function $\sigma(u)$ with points of increase λ_j $(j = 1, 2, 3, ...)$ and discontinuities μ_j $(j = 1, 2, 3, ...)$ and we construct the space L_σ^2 together with the sequence of its orthonormal polynomials $\{P_k(u)\}_0^\infty$. Then we introduce the functions $\kappa_k(u) \in L_\sigma^2$ $(k = 1, 2, 3, ...)$ determined by the equations

$$\kappa_k(\lambda_j) = \begin{cases} \dfrac{1}{\sqrt{\mu_k}} & (j = k), \\ 0 & (j \neq k). \end{cases}$$

These functions are normalized and any pair of them is orthogonal:

$$\int_{-\infty}^{\infty} \kappa_i(u)\kappa_k(u) \, d\sigma(u) = \delta_{ik}.$$

If we take an arbitrary complex sequence $\{\gamma_k\}_1^\infty$, for which

$$\sum_1^\infty |\gamma_k|^2 < \infty, \qquad \qquad ...[4.33]$$

then with the aid of the equation

$$\underset{\sigma(u)}{\text{l.i.m.}} \sum_1^n \gamma_k \kappa_k(u) = g(u)$$

we can define a certain function $g(u) \in L_\sigma^2$. We shall need only one such function, namely that which corresponds to the sequence

$$\gamma_k = \frac{q(z)}{\sqrt{\mu_k} q'(\lambda_k)(z - \lambda_k)} \quad (k = 1, 2, 3, ...)$$

for some non-real z. For this sequence the inequality [4.33] is satisfied by virtue of condition [4.31]. The Fourier coefficients of the polynomials $P_j(u)$ for this special function $g(u)$ are

$$\int_{-\infty}^{\infty} g(u)P_j(u) \, d\sigma(u) = (g, P_j)_\sigma = \lim_{n \to \infty} (g_n, P_j)_\sigma,$$

where

$$g_n = g_n(u) = \sum_1^n \gamma_k \kappa_k(u).$$

Since

$$(\kappa_k, P_j)_\sigma = \int_{-\infty}^{\infty} \kappa_k(u)P_j(u) \, d\sigma(u) = \frac{1}{\sqrt{\mu_k}} P_j(\lambda_k)\mu_k,$$

we have

$$(g, P_j)_\sigma = \lim_{n \to \infty} \sum_{k=1}^{n} \gamma_k \sqrt{\mu_k} P_j(\lambda_k) = q(z) \sum_{k=1}^{\infty} \frac{P_j(\lambda_k)}{q'(\lambda_k)(z - \lambda_k)}.$$

From [4.28a] we find that

$$(g, P_j)_\sigma = P_j(z) \quad (j = 0, 1, 2, \ldots).$$

Now we write down the Bessel inequality for $g(u)$:

$$\sum_{0}^{\infty} | P_j(z) |^2 \leqq (g, g)_\sigma < \infty.$$

This inequality shows that the moment problem is indeterminate, which is what had to be proved.

2. DEFINITION 2. We shall call the sequence $\{\lambda_j\}_1^\infty$ *a canonical sequence of nodes* and the sequence of numbers $\mu_j > 0$ ($j = 1, 2, 3, \ldots$), $\sum_1^\infty \mu_j = 1$, belonging to this sequence *the sequence of masses*, if there exists an indeterminate moment problem of which some canonical solution has the sequence $\{\lambda_j\}_1^\infty$ for all its points of increase and the numbers μ_j correspondingly for its discontinuities.

THEOREM. *Any canonical sequence of nodes* $\{\lambda_j\}_1^\infty$ *coincides with the aggregate of the zeros of some function* $q(z)$ *of the class* $\mathfrak{A}$. *If* $\{\mu_j\}_1^\infty$ *is the sequence of masses belonging to the sequence* $\{\lambda_j\}_1^\infty$, *then*

$$\sum_{1}^{\infty} \frac{1}{\mu_j(1 + \lambda_j^2)[q'(\lambda_j)]^2} < \infty, \qquad \ldots[4.31]$$

and

$$\sum_{1}^{\infty} \frac{1}{\mu_j[q'(\lambda_j)]^2} = \infty. \qquad \ldots[4.34]$$

Proof. We begin with the fact that there exists a real entire function of zero degree, $q(z)$ say, all of whose zeros are simple and coincident with the numbers λ_j ($j = 1, 2, 3, \ldots$). Apart from a constant real factor this function $q(z)$ is equal to the denominator of the fraction

$$w(z) = -\frac{A(z)t - C(z)}{B(z)t - D(z)}$$

for some constant real value of t, where

$$\left\| \begin{array}{cc} A(z) & C(z) \\ B(z) & D(z) \end{array} \right\|$$

is the Nevanlinna matrix belonging to the indeterminate moment problem which is generated by the moments

$$s_m = \sum_{j=1}^{\infty} \mu_j \lambda_j^m \quad (m = 0, 1, 2, \ldots). \qquad \ldots[4.32]$$

For definiteness we assume that

$$q(z) = D(z).$$

In this case we have

$$-B(\lambda_j)C(\lambda_j) = 1, \quad \mu_j = \frac{C(\lambda_j)}{q'(\lambda_j)} = -\frac{1}{B(\lambda_j)q'(\lambda_j)}$$

and (see Addenda and Problems to Chapter 3, example 12)

$$\operatorname{Im} \frac{B(z)}{q(z)} : \operatorname{Im} z > 0 \quad (\operatorname{Im} z \neq 0),$$

i.e. $B(z)/q(z) \in N$ whence it follows that $B(z)/q(z)$ permits the absolutely convergent expansion

$$\frac{B(z)}{q(z)} = c_0 + c_1 z - \frac{A}{z} + z \sum_{\lambda_j \neq 0} \frac{A_j}{(\lambda_j - z)\lambda_j}, \qquad \ldots[4.35]$$

where

$$c_1 \geq 0, \quad A \geq 0, \quad A_j = -\frac{B(\lambda_j)}{q'(\lambda_j)} = \frac{1}{\mu_j [q'(\lambda_j)]^2} \quad (j = 1, 2, 3, \ldots).$$

From the fact that the expansion $[4.35]$ is absolutely convergent it follows that the series

$$\sum_{\lambda_j \neq 0} \frac{1}{\lambda_j^2 \mu_j [q'(\lambda_j)]^2}$$

is convergent i.e. that relation $[4.31]$ holds.

Further, applying Cauchy's inequality we find that for any integer $m \geq 0$ we have

$$\sum_{j=1}^{\infty} \left| \frac{\lambda_j^m}{q'(\lambda_j)} \right| = \sum_{j=1}^{\infty} \left| \frac{1}{\sqrt{\mu_j}\sqrt{1+\lambda_j^2} q'(\lambda_j)} \right| \cdot | \sqrt{\mu_j} \lambda_j^m \sqrt{1+\lambda_j^2} |$$

$$\leq \sqrt{\sum_{j=1}^{\infty} \frac{1}{\mu_j(1+\lambda_j^2)[q'(\lambda_j)]^2}} \sqrt{\sum_{j=1}^{\infty} \mu_j(\lambda_j^{2m} + \lambda_j^{2m+2})}.$$

Therefore all series $[4.29]$ converge absolutely.

The absolute convergence of the series

$$\sum_{1}^{\infty} \frac{1}{q'(\lambda_j)(z-\lambda_j)}$$

is now evident.

Going over the proof of the remaining statements we note that for $\pm y \to \infty$ the quantity $q(iy)$ must increase more rapidly than any polynomial, since if it did not the function $q(z)$, being a function of minimal exponential type, would itself be a polynomial.

We take an arbitrary polynomial $S(z)$ and consider the difference

$$\phi(z) = S(z) - q(z) \sum_{1}^{\infty} \frac{S(\lambda_j)}{q'(\lambda_j)(z-\lambda_j)},$$

which is an entire function of minimal exponential type. We have that

$$\phi(\lambda_j) = 0 \quad (j = 1, 2, 3, \ldots)$$

and therefore

$$\frac{\phi(z)}{q(z)} = \psi(z)$$

is also an entire function; it is also of minimal exponential type. But by virtue of the remark just made we have

$$\lim_{+y \to \infty} \psi(iy) = 0.$$

Therefore $\psi(z) \equiv 0$ i.e. for any polynomial $S(z)$ the equation

$$S(z) = q(z) \sum_{1}^{\infty} \frac{S(\lambda_j)}{q'(\lambda_j)(z-\lambda_j)}$$

is true. Taking $S(z) = 1$ we obtain the expansion [4.28] and this gives us the proof that $q(z) \in \mathfrak{A}$.

It remains to prove the relation [4.34]. To do this we introduce the new masses

$$\mu_j^* = \frac{\mu_j}{1+\lambda_j^2} \quad (j = 1, 2, 3, \ldots)$$

and determine their moments

$$s_m^* = \sum_{j=1}^{\infty} \mu_j^* \lambda_j^m \quad (m = 0, 1, 2, \ldots).$$

It follows from this definition that

$$s_1^* + is_0^* = \sum_{k=1}^{\infty} \frac{\mu_k}{\lambda_k - i} \quad s_m^* + s_{m+2}^* = s_m \quad (m = 0, 1, 2, \ldots).$$

Therefore, to any solution $\sigma^*(u)$ of the new moment problem

$$\int_{-\infty}^{\infty} u^m \, d\sigma^*(u) = s_m^* \quad (m = 0, 1, 2, \ldots)$$

the formula

$$\sigma(u) = \int_{-\infty}^{u} (1 + u^2) \, d\sigma^*(u)$$

relates such a solution of the old moment problem, for which

$$\int_{-\infty}^{\infty} \frac{d\sigma(u)}{u - i} \equiv w(i) = \sum_{k=1}^{\infty} \frac{\mu_k}{\lambda_k - i}.$$

But the system $\{\lambda_j, \mu_j\}_1^\infty$ corresponds to the canonical solution of the old moment problem. Therefore the point $w(i)$ lies on the periphery of the circle $K_\infty(i)$ and therefore the new moment problem is determinate. Relation [4.34] is now a consequence of Lemma 2.

Thus the theorem is proved.

COROLLARY 1. *Let* A *be a simple closed symmetric operator which for a certain base* $\{e_k\}_0^\infty$ *can be represented by a* $\mathscr{J}$-*matrix of type* C:

$$Ae_k = b_k e_{k+1} + a_k e_k + b_{k-1} e_{k-1} \quad (k = 0, 1, 2, \ldots; \ b_{-1} = 0).$$

Assume further that $\mathring{A}$ *is some self-adjoint extension of the operator* A *not involving an extension of the space and that* $\mathring{E}_u$ *is a resolution of unity for the operator* $\mathring{A}$.

In such a case the spectrum of the operator $\mathring{A}$ *consists of the aggregate of zeros* λ_j $(j = 1, 2, 3, \ldots)$ *of some function* $q(z) \in \mathfrak{A}$ *and the discontinuities*

$$\mu_j = \sigma(\lambda_j + 0) - \sigma(\lambda_j) \quad (j = 1, 2, 3, \ldots)$$

of the function $\sigma(u) = (\mathring{E}_u e_0, e_0)$ *are such that the relations* [4.31] *and* [4.34] *are valid.*

3. THEOREM 2. *Let* $\mathring{A}$ *be a self-adjoint operator whose spectrum is formed by the zeros* λ_j $(j = 1, 2, 3, \ldots)$ *of the function* $q(z) \in \mathfrak{A}$.

In this case every normalized sequence of positive numbers μ_j $(j = 1, 2, 3, \ldots)$, *for which the relations* [4.31] *and* [4.34] *are valid and the series*

$$\sum_{j=1}^{\infty} \mu_j \lambda_j^{2m} \quad (m = 0, 1, 2, \ldots)$$

converge can be put in correspondence with a base $\{e_k\}_0^\infty$ and a $\mathscr{J}$-matrix of type C which generate a symmetric operator $A \subset \mathring{A}$ with a deficiency index (1, 1).

Proof. We denote by κ_j $(j = 1, 2, 3, ...)$ the normalized characteristic vectors of the operator $\mathring{A}$:

$$\mathring{A}\kappa_j = \lambda_j\kappa_j \quad (j = 1, 2, 3, ...).$$

These span the space H, in which the operator $\mathring{A}$ acts and form an orthonormal base in H. Just as in example 1, we introduce the sectionally constant function $\sigma(u)$ with discontinuities

$$\sigma(\lambda_j+0) - \sigma(\lambda_j-0) = \mu_j \quad (j = 1, 2, 3, ...).$$

We map the space H on to L_σ^2 by relating to the vector $\kappa_k \in H$ the function $\kappa_k(u) \in L_\sigma^2$, determined by the equation

$$\kappa_k(\lambda_j) = \frac{1}{\sqrt{\mu_k}}\delta_{kj}.$$

To the operator $\mathring{A}$ in H there corresponds in L_σ^2 the operator of multiplying by the independent variable and we shall also call this operator $\mathring{A}$. Our problem consists in constructing an operator $A \subset \mathring{A}$ which satisfies the conditions of the theorem. This construction, and the whole discussion, can be performed not in H but in L_σ^2.

In the first place we introduce the element $g(u) \in L_\sigma^2$ with components

$$(g(u), \kappa_k(u))_\sigma = \frac{\Omega_k}{\lambda_k - i} \quad (k = 1, 2, 3, ...),$$

where

$$\Omega_k = \frac{1}{\sqrt{\mu_k}q'(\lambda_k)}.$$

The convergence of the series

$$\sum_1^\infty \frac{\Omega_k^2}{1+\lambda_k^2}$$

is guaranteed by condition [4.31].

For our D_A we take the aggregate of all functions of the form

$$f(u) = (\mathring{A}+iE)^{-1}h(u) = \frac{h(u)}{u+i}, \qquad ...[4.36]$$

where $h(u) \in L_\sigma^2$ and

$$(h(u), g(u))_\sigma = 0. \qquad ...[4.36a]$$

Hence it follows directly that if $f(u) \in D_A$, then

$$\sum_1^\infty (f(u), \kappa_k(u))_\sigma \Omega_k = 0.$$

Since any function of the form $[4.36]$ belongs to $D_{\hat{A}}$ we have $D_A \subset D_{\hat{A}}$. To define the operator A we put

$$Af(u) = \hat{A}f(u) \quad [f(u) \in D_A].$$

We prove that D_A is dense in L_σ^2, whence it will follow that the operator A is symmetric. Assume that a function $\phi(u) \neq 0$ from L_σ^2 is orthogonal to D_A. This means that for any function $h(u) \in L_\sigma^2$ satisfying the equation $[4.36a]$ we have:

$$\left(\phi(u), \frac{h(u)}{u+i} \right)_\sigma = 0$$

or

$$\left(\frac{\phi(u)}{u-i}, h(u) \right)_\sigma = 0,$$

whence it follows that

$$\frac{\phi(u)}{u-i} = \gamma g(u),$$

where $\gamma \neq 0$. Therefore

$$(\phi(u), \kappa_k(u))_\sigma = \gamma(\lambda_k - i)(g(u), \kappa_k(u))_\sigma = \gamma \Omega_k$$

and thus

$$\sum_1^\infty \Omega_k^2 < \infty,$$

and this contradicts equation $[4.34]$.

Now we prove that the operator A has deficiency index $(1, 1)$. To do this we note that the sub-space $\mathfrak{N}_i$ consists of all functions $\psi(u) \in L_\sigma^2$, for which the equation

$$((A+iE)f(u), \psi(u))_\sigma = 0. \qquad \qquad \dots [4.37]$$

is satisfied with any function $f(u) \in D_A$. Since

$$f(u) = \frac{h(u)}{u+i},$$

where

$$h(u) \in L^2_\sigma, \quad (h(u), g(u))_\sigma = 0, \qquad \qquad ...[4.38]$$

equation [4.37] shows that [4.38] implies

$$(h(u), \psi(u))_\sigma = 0.$$

Therefore

$$\psi(u) = \gamma g(u),$$

where $\gamma \neq 0$, and so the sub-space $\mathfrak{N}_i$ is one-dimensional and is generated by the function $g(u)$. Similarly we find that the sub-space $\mathfrak{N}_{-i}$ is also one-dimensional and is generated by the function $\overline{g(u)}$.

Thus the symmetric operator $A \subset \dot{A}$ is constructed.

Now one must construct an orthonormal base $\{e_k(u)\}_0^\infty$ in L^2_σ in such a way that for this base the matrix of the operator A is a $\mathscr{J}$-matrix.

We maintain that such a base is formed by the polynomials $P_k(u)$ ($k = 0, 1, 2, ...$) which are orthonormal relative to the sequence [4.32]. Indeed these polynomials form a base in L^2_σ since the operator $\dot{A}$ is self-adjoint and therefore the set of all polynomials is dense in L^2_σ.

Together with the orthonormal polynomials a certain $\mathscr{J}$-matrix is defined.

We must only prove that every polynomial belongs to D_A. Assume that $S(u)$ is some polynomial. We have to prove that it belongs to D_A which means that we must show that

$$((u + i)S(u), g(u))_\sigma = 0$$

or

$$\sum_1^\infty \frac{S(\lambda_j)}{q'(\lambda_j)} = 0.$$

But these equations do indeed hold as was shown in example 1 (see [4.29a].

Thus Theorem 2 is proved.

COROLLARY 2. *The necessary condition established by Theorem 1 for a sequence of canonical nodes and the sequence of masses belonging to it, is also sufficient, namely: the set $\{\lambda_j\}_1^\infty$ of all zeros of a function $q(z) \in \mathfrak{A}$ is a canonical sequence of nodes and any sequence of positive numbers μ_j ($j = 1, 2, 3, ...$) for which the relations [4.31] and [4.34] are valid and all series*

$$\sum_{j=1}^\infty \mu_j \lambda_j^{2m} \quad (m = 0, 1, 2, ...)$$

converge is a sequence of masses belonging to the sequence $\{\lambda_j\}_1^\infty$.

The function $q(z)$ is an element of the Nevanlinna matrix generated by an indeterminate moment problem if and only if $q(z) \in \mathfrak{A}$.

4. Let A be a simple closed symmetric operator in H with deficiency index $(1, 1)$.

Further let $\hat{A}$ be any self-adjoint extension of the operator A which does not involve extending H. In order that the operator A should be representable with the aid of matrices of type C it is necessary that the operator $\hat{A}$ should have only a discrete spectrum $\{\lambda_j\}_1^\infty$, this spectrum being formed by the zeros of a function of class $\mathfrak{A}$. Let this condition be satisfied and assume that $\{\lambda_j\}_1^\infty$ is a complete orthonormal system of characteristic vectors for the operator $\hat{A}$:

$$\hat{A}\kappa_j = \lambda_j\kappa_j \quad (j = 1, 2, 3, \ldots).$$

The domain of the operator A as a restriction of the operator $\hat{A}$ is the set of all vectors of the form

$$f = (\hat{A} + iE)^{-1}h,$$

where h is an arbitrary element of H satisfying the condition

$$\sum_{k=1}^\infty (h, \kappa_k)\frac{\Omega_k}{\lambda_k + i} = 0,$$

where all the $\Omega_k \neq 0$ and satisfy the condition:

$$\sum_1^\infty |\Omega_k|^2 = \infty, \quad \sum_1^\infty \frac{|\Omega_k|^2}{1 + \lambda_k^2} < \infty.$$

The operator A is representable by the matrix of C if and only if

$$\sum_{k=1}^\infty \frac{\lambda_k^{2m}}{|\Omega_k|^2 |q'(\lambda_k)|^2} < \infty \quad (m = 0, 1, 2, \ldots)$$

(HAMBURGER[4]).

The essence of the proof is the formula which represents the resolvents of all self-adjoint extensions of the given symmetric operator A with deficiency index (1.1). This formula is reminiscent of the Nevanlinna formula and was first obtained by HAMBURGER[4]. Independently, this formula and also some general formulae for resolvents, were obtained by Krein (see e.g. AKHIEZER and GLAZMAN[1]).

5. Assume that an infinite sequence of numbers

$$\lambda_1 < \lambda_2 < \lambda_3 < \ldots$$

is given and assume that for any $r > 0$

$$n(r) = \sum_{\lambda_k \leq r} 1.$$

If there exist numbers $\gamma > 0$, $d > 0$, $0 < \rho < \frac{1}{2}$ such that
a) for $r \to \infty$ the asymptotic equation

$$n(r) \sim \gamma r^\rho$$

holds and

b) $\inf_k \left[\lambda_k^\rho - \lambda_{k-1}^\rho \right] \geqq d$,

then the sequence of numbers $\{\lambda_k\}_1^\infty$ is a canonical sequence of nodes.

In the form here given the theorem represents a correction to a proposition given by HAMBURGER[4]. The error made by Hamburger consists in the fact that he omitted condition b); it was noted by Levin who also found the correct formulation. This remark also applies to the following example.

6. Assume that two positive sequences

$$\lambda_1 < \lambda_2 < \lambda_3 < \ldots; \quad \lambda_0' < \lambda_1' < \lambda_2' < \ldots$$

are given and assume that the functions

$$n(r) = \sum_{\lambda_k \leqq r} 1, \quad n'(r) = \sum_{\lambda_k' \leqq r} 1$$

are subject to the asymptotic equations

$$n(r) = \gamma r^\rho, \quad n'(r) \sim \gamma' r^{\rho'} \quad (r \to \infty),$$

where γ, γ', ρ, ρ', are some positive numbers. Assume in addition that there exists a number $d > 0$, such that

$$\inf_k \left[\lambda_k^\rho - \lambda_{k-1}^\rho \right] \geqq d, \quad \inf_k \left[\lambda_k'^{\rho'} - \lambda_{k-1}'^{\rho'} \right] \geqq d.$$

In such a case the sequence

$$\ldots < -\lambda_n' < -\lambda_{n-1}' < \ldots < -\lambda_0' < \lambda_1 < \lambda_2 < \ldots < \lambda_n < \ldots$$

must certainly be a canonical sequence of nodes, provided one of the following conditions is satisfied:

a) $\rho < \dfrac{1}{2}; \quad \rho' < \dfrac{1}{2},$

b) $\rho = \rho', \quad \gamma \neq \gamma', \quad \tan^2 \dfrac{\pi\rho}{2} < \dfrac{\gamma + \gamma'}{|\gamma - \gamma'|},$

c) $\rho = \rho', \quad \gamma = \gamma', \quad 0 < \rho < 1.$

7. A complex moment problem over the whole plane G can be formulated as follows: it is required to find a non-negative measure $\omega(A)$ such that

$$c_{\alpha\beta} = \int_G z^\alpha \bar{z}^\beta \, d\omega(z) \quad (\alpha, \beta = 0, 1, 2, \ldots), \qquad \ldots [4.39]$$

where the $c_{\alpha\beta}$ are given; here $\omega(z) = \omega(A_z)$, where A_z is the set of all points ζ for which

$$\text{Re } \zeta \leqq \text{Re } z, \quad \text{Im } \zeta \leqq \text{Im } z.$$

If one requires in addition that the spectrum of $\omega(A)$ does not reduce to a finite system of points, it is necessary for the problem to be soluble that the inequalities

$$\begin{vmatrix} c_{00} & c_{01} & \cdots & c_{0n} \\ c_{10} & c_{11} & \cdots & c_{1n} \\ \cdot & \cdot & \cdots & \cdot \\ \cdot & \cdot & \cdots & \cdot \\ c_{n0} & c_{n1} & \cdots & c_{nn} \end{vmatrix} > 0 \quad (n = 0, 1, 2, \ldots).$$

be satisfied. If these inequalities are satisfied, then there is uniquely associated with the moment problem a certain normal operator N in a Hilbert space H and a certain element $f \in$ H such that

$$c_{\alpha\beta} = (N^\alpha \bar{N}^\beta f, f) \quad (\alpha, \beta = 0, 1, 2, \ldots).$$

The normal property of the operator N means that: α) D_N is dense in H, β) $D_N \subseteq D_{N^*}$ and $\| Ng \| = \| \bar{N}g \| \ (g \in D_N)$, where $\bar{N}$ is the part of N^* lying in D_N.

For $D_N = D_{N^*}$ the normal operator is called hypermaximal.

If N is a normal operator associated with the moment problem in question the deficiency index of each of the symmetric operators

$$\frac{1}{2}(N + \bar{N}), \quad \frac{1}{2i}(N - \bar{N})$$

is $(0, 0)$ or $(1, 1)$.

The moment problem $[4.39]$ is soluble if and only if the operator N can be extended to a hypermaximal normal operator. The determinateness or indeterminateness of the moment problem depends on whether this extension is unique or not.

The results just listed are due to KILPI[1].

8. The " pure operator approach " to the Nevanlinna-Pick interpolation problem (NAGY and KORANYI[1, 2]).[†]

† See also KORANYI[1].

A point set Z is given in the unit circle and on it a function $\Phi(\zeta)$ is defined; it is known that for any positive integer n, any $\zeta_\alpha \in Z$ and any complex ρ_α the inequality

$$\sum_{\alpha,\,\beta=1}^{n} \frac{\Phi(\zeta_\alpha)+\overline{\Phi(\zeta_\beta)}}{1-\zeta_\alpha\bar{\zeta}_\beta}\rho_\alpha\bar{\rho}_\beta \geqq 0. \qquad \qquad ...[4.40]$$

holds. It is required to prove that there exists a function $F(z) \in C$ for which

$$F(\zeta) = \Phi(\zeta) \quad (\zeta \in Z). \qquad \qquad ...[4.41]$$

The original proof due to the Hungarian mathematicians is based on the construction of a Hilbert space H and in it of a certain unitary operator U such that for any $\zeta \in Z$ one has

$$\Phi(\zeta) = i\mu+\frac{1}{2}\left(\frac{E+\zeta U}{E-\zeta U}e_0,\, e_0\right), \qquad \qquad ...[4.42]$$

where e_0 is some element of the space H and μ is some real constant; having done this it is sufficient to put

$$F(z) = i\mu+\frac{1}{2}\left(\frac{E+zU}{E-zU}e_0,\, e_0\right) \quad (\,|\,z\,|\, < 1).$$

Indeed the function $F(z)$ evidently satisfies condition [4.41], and its analyticity in the circle $|\,z\,| < 1$ follows from the expansion

$$F(z) = \frac{1}{2}(e_0,\, e_0)+i\mu+\sum_{1}^{\infty}(U^k e_0,\, e_0)z^k,$$

since

$$|\,(U^k e_0,\, e_0)\,| \leqq (e_0,\, e_0) \quad (k = 1, 2, 3, ...).$$

Finally, if one puts

$$e_0 = (E-zU)g \quad (g = g_z),$$

then

$$F(z) = i\mu+\frac{1}{2}((E+zU)g,$$

$$(E-zU)g) = i\mu+\frac{1}{2}(g,\, g)+\frac{1}{2}(Ug,\, g)z-\frac{1}{2}(g,\, Ug)\bar{z}-\frac{|\,z\,|^2}{2}(Ug,\, Ug),$$

whence

$$\mathrm{Re}\,F(z) = \frac{1-|\,z\,|^2}{2}(g,\, g) > 0 \quad (\,|\,z\,| < 1)$$

and so $F(z) \in C$.

For simplicity we assume that the hermitian form appearing in equation [4.40] is always positive and we also assume, without restricting generality, that the point $z = 0$ belongs to Z. To construct the Hilbert space we relate to every point $\zeta \in Z$ the symbol e_ζ and consider the linear manifold G of all possible forms

$$\sum \rho_\alpha e_{\zeta_\alpha}$$

with any complex ρ_α. Then we perform the metrization of the manifold G with the aid of the kernel

$$K(s, t) = \frac{\Phi(s) + \overline{\Phi(t)}}{1 + s\bar{t}} \quad (s, t \in Z),$$

defining the scalar product of the elements

$$f = \sum_l \xi_l e_{s_l}, \quad g = \sum_m \eta_m e_{t_m}$$

by the formula

$$(f, g) = \sum_{l, m} K(s_l, t_m) \xi_l \bar{\eta}_m.$$

This definition satisfies all requirements necessary for a scalar product. Finally we close the manifold G with respect to the norm just introduced. In this way we obtain a " preliminary " Hilbert space H. In it we construct the linear operator V, which is defined over the elements $e_\zeta (\zeta \neq 0)$ by the equation

$$V e_\zeta = \frac{1}{\zeta}(e_\zeta - e_0). \qquad \ldots[4.43]$$

Evidently the operator V is not defined on the whole Hilbert space H. If $s, t \in Z$ and $s \neq 0$, $t \neq 0$, then

$$(V e_s, V e_t) = \frac{1}{s\bar{t}}(e_s - e_0, e_t - e_0)$$

$$= \frac{1}{s\bar{t}}[(e_s, e_t) - (e_s, e_0) - (e_0, e_t) + (e_0, e_0)]$$

$$= \frac{1}{s\bar{t}}\left[\frac{\Phi(s) + \overline{\Phi(t)}}{1 - s\bar{t}} - \Phi(s) - \overline{\Phi(0)} - \Phi(0) - \overline{\Phi(t)} + \Phi(0) + \overline{\Phi(0)}\right]$$

$$= \frac{\Phi(s) + \overline{\Phi(t)}}{1 - s\bar{t}} = (e_s, e_t).$$

Hence it follows that the operator V is isometric. We extend it to a unitary operator U, possibly by expanding into a wider Hilbert space $\mathscr{H}$ which will be the final Hilbert space.

Now we write down equation $[4.43]$ in the form

$$(E - \zeta U)e_\zeta = e_0, \qquad \dots [4.44]$$

which is now true for all $\zeta \in Z$ (including $\zeta = 0$).

Further, we write down the identity

$$\frac{1}{2}\frac{E + \zeta U}{E - \zeta U}e_0 = (E - \zeta U)^{-1}e_0 - \frac{1}{2}e_0$$

and by virtue of $[4.44]$ we have

$$\frac{1}{2}\frac{E + \zeta U}{E - \zeta U}e_0 = e_\zeta - \frac{1}{2}e_0.$$

Taking the scalar product of both sides with e_0 and using the definition of the scalar product we obtain

$$\frac{1}{2}\left(\frac{E + \zeta U}{E - \zeta U}e_0, e_0\right) = \Phi(\zeta) - i \operatorname{Im} \Phi(0) \quad (\zeta \in Z).$$

This is just equation $[4.42]$. Thus the proof is complete.

Chapter 5

TRIGONOMETRIC AND CONTINUOUS ANALOGUES

The beginning of the present chapter follows on directly after Chapter 3 and is devoted to the trigonometric moment problem which, in essence, is nothing other than the Carathéodory coefficient problem. Further we consider in a somewhat condensed form trigonometric analogues of orthogonal polynomials and of some other objects which have been introduced in connection with the power moment problem.

Subsequently we study the problem of integral representation of hermitian-positive functions, this problem being the continuous analogue of the trigonometric moment problem.

The end of the chapter is devoted to certain classes of monotonic functions, the integral representation of which can be considered as a continuous analogue of the power moment problem.

It should be noted that these discussions do not by far exhaust the various continuous analogues both of the trigonometric and of the power moment problem. Certain further results related to these matters and references to the literature are given in the addenda to the present chapter.

1. The trigonometric moment problem

1.1. We begin with the Carathéodory coefficient problem: find the conditions which are necessary and sufficient in order that the function

$$F(z) = c + c_1 z + c_2 z^2 + \dots + c_n z^n + \dots \qquad \dots [5.1]$$

should belong to the class C.

As was remarked in Section **1.1** of Chapter 3, this problem reduces to that of Schur coefficients and therefore can be solved by the method of successive linear fractional transformations. However there is no need to resort to the general theory here because the required conditions may be very simply formulated and then directly verified.

THEOREM **5.1.1** (CARATHÉODORY[1], TOEPLITZ[1]). *In order that the* *function* [5.1] *should belong to the class* C *it is necessary and sufficient that the Toeplitz form*

$$\sum_{\alpha, \beta=0}^{n} c_{\alpha-\beta}\xi_\alpha\bar{\xi}_\beta, \qquad \ldots[5.2]$$

where $c_0 = c+\bar{c}$, $c_{-k} = \bar{c}_k$, *should be non-negative for every n.*

We note to start with that the equation $c_{-k} = \bar{c}_k$, and also the in-equalities $|c_k| \leq c_0$ ($k = 0, 1, 2, \ldots$) are consequences of the fact that the forms [5.2] are non-negative. Indeed, put $n = k$, $\xi_1 = \xi_2 = \ldots = \xi_{k-1} = 0$, $\xi_k = 1$. We then obtain the inequality

$$c_0 |\xi_0|^2 + c_k\bar{\xi}_0 + c_{-k}\xi_0 + c_0 \geq 0,$$

Hence it follows that $c_{-k} = \bar{c}_k$. Then we put $\xi_0 = te^{i\vartheta}$ where $\vartheta = \arg c_k$, and t is real. Our inequality now takes on the form

$$c_0 t^2 + 2|c_k| t + c_0 \geq 0 \quad (-\infty < t < \infty),$$

whence it follows that $|c_k| \leq c_0$.

Now we come to the proof of the theorem.

The necessity of the condition is proved most simply by use of the Riesz-Herglotz integral representation

$$F(z) = \frac{c-\bar{c}}{2} + \frac{1}{2\pi}\int_{-\pi}^{\pi} \frac{e^{i\theta}+z}{e^{i\theta}-z} \, d\omega(\theta),$$

where $\omega(\theta)$ $(-\pi \leq \theta \leq \pi)$ is a non-decreasing function. Indeed by virtue of this representation we have

$$F(z) = \frac{c-\bar{c}}{2} + \frac{1}{2\pi}\int_{-\pi}^{\pi} d\omega(\theta) + \sum_{k=1}^{\infty} z^k \frac{1}{\pi}\int_{-\pi}^{\pi} e^{-ik\theta} \, d\omega(\theta),$$

whence

$$c_k = \frac{1}{\pi}\int_{-\pi}^{\pi} e^{-ik\theta} \, d\omega(\theta) \quad (k = 0, 1, 2, \ldots).$$

Therefore

$$\sum_{\alpha, \beta=0}^{n} c_{\alpha-\beta}\xi_\alpha\bar{\xi}_\beta = \frac{1}{\pi}\int_{-\pi}^{\pi} \left| \sum_{k=0}^{n} \xi_k e^{-ik\theta} \right|^2 d\omega(\theta) \geq 0. \qquad \ldots[5.3]$$

The proof of sufficiency is equally simple. Indeed, if all the forms [5.2] are non-negative, then as we noted above $|c_k| \leq c_0 (k = 1, 2, 3, \ldots)$

and therefore the function $[5.1]$ is regular in the circle $|z| < 1$. Since we then have†

$$\frac{2 \operatorname{Re} F(z)}{1-|z|^2} = \frac{F(z)+\overline{F(z)}}{1-z\bar{z}} = \sum_{m=0}^{\infty} z^m \bar{z}^m \left\{ \sum_{k=0}^{\infty} c_k z^k + \sum_{k=1}^{\infty} c_{-k} \bar{z}^k \right\}$$

$$= \sum_{m=0}^{\infty} \sum_{k=0}^{\infty} c_k z^{m+k} \bar{z}^m + \sum_{m=0}^{\infty} \sum_{k=1}^{\infty} c_{-k} z^m \bar{z}^{m+k}$$

$$= \sum_{\beta=0}^{\infty} \sum_{\alpha=m}^{\infty} c_{\alpha-\beta} z^\alpha \bar{z}^\beta + \sum_{\alpha=0}^{\infty} \sum_{\beta=m+1}^{\infty} c_{\alpha-\beta} z^\alpha \bar{z}^\beta$$

$$= \sum_{\alpha,\beta=0}^{\infty} c_{\alpha-\beta} z^\alpha \bar{z}^\beta = \lim_{n\to\infty} \sum_{\alpha,\beta=0}^{n} c_{\alpha-\beta} z^\alpha \bar{z}^\beta \geqq 0,$$

it follows that $F(z)$ belongs to the class C and the proof is complete.

1.2. The theorem **5.1.1** which we have just proved expresses the criterion for the so-called trigonometric moment problem to be soluble; this problem consists of the following: *given an infinite sequence of numbers* $\{c_k\}_{-\infty}^{\infty}$; *it is required to find a non-decreasing function* $\sigma(\theta)$ $(-\pi \leqq \theta \leqq \pi)$ *such that*

$$c_k = \frac{1}{2\pi} \int_{-\pi}^{\pi} e^{ik\theta} d\sigma(\theta) \quad (\pm k = 0, 1, 2, \ldots). \qquad \ldots[5.4]$$

Here it is useful to assume the following definition:

DEFINITION **5.1.2.** We call an infinite sequence of numbers $\{c_k\}_{-\infty}^{\infty}$ *non-negative (positive) relative to the circumference of a circle*, if for any n the Toeplitz form $[5.2]$ is non-negative (positive).

THEOREM **5.1.2** (F. RIESZ[1], HERGLOTZ[1]). a) *In order that the trigonometric moment problem* $[5.4]$ *be soluble it is necessary and sufficient that the sequence* $\{c_k\}_{-\infty}^{\infty}$ *should be non-negative relative to the circumference of the circle.* b) *The trigonometric moment problem cannot be indeterminate.* c) *The solution* $\sigma(\theta)$ *has an infinite set of points of increase if and only if the sequence* $\{c_k\}_{-\infty}^{\infty}$ *is positive.*

To prove statement b) we perform an integration by parts in the equations $[5.4]$. These equations then take on the form

$$c_k = (-1)^k \frac{\sigma(\pi)-\sigma(-\pi)}{2\pi} - \frac{ik}{2\pi} \int_{-\pi}^{\pi} \sigma(\theta) e^{ik\theta} d\theta \quad (\pm k = 0, 1, 2, \ldots)$$

† This argument is also applicable in the continuous case (see Addenda and Problems to Chapter 5, example 9).

and they show that all the Fourier coefficients of the solution $\sigma(\theta)$ of equation [5.4], except the zero-th coefficient, are known:

$$\frac{1}{2\pi}\int_{-\pi}^{\pi} \sigma(\theta)e^{ik\theta}\,d\sigma = -\frac{c_k-(-1)^k c_0}{ik} \quad (\pm k = 1, 2, 3, ...).$$

Therefore $\sigma(\theta)$ is determined by the equations [5.4], apart from an additive constant.

Statement c) is easy to derive with the aid of relation [5.3].

1.3. We agree to give the name *quasi-polynomials* to expressions of the form

$$\sum_{k=-m}^{n} A_k z^k,$$

where $m \geq 0$, $n \geq 0$. In particular for $m = 0$ we obtain polynomials proper.

Taking any sequence $\{c_k\}_{-\infty}^{\infty}$ we can define a functional $\mathfrak{C}$ linear in the algebraic sense in the space of all quasi-polynomials by putting

$$\mathfrak{C}\{z^k\} = c_k \quad (\pm k = 0, 1, 2, ...).$$

The non-negative character of the sequence $\{c_k\}_{-\infty}^{\infty}$ with respect to the circumference of the circle implies an important property of the functional $\mathfrak{C}$, namely: *if the quasi-polynomial $T(z)$ satisfies the condition*

$$T(e^{i\theta}) \geq 0 \quad (-\pi \leq \theta \leq \pi)$$

(more briefly: *if it is non-negative on the unit circle*) *then the functional $\mathfrak{C}$ associates with it a number*

$$\mathfrak{C}\{T(z)\} \geq 0.$$

To prove this one must bear in mind that any quasi-polynomial $T(z)$ which is non-negative on the unit circle permits the representation

$$T(e^{i\theta}) = \left| \sum_{k=0}^{n} \xi_k e^{ik\theta} \right|^2.$$

It is not difficult to see what further statements can be made if the sequence $\{c_k\}_{-\infty}^{\infty}$ is positive.

To solve the trigonometric problem means to realize the functional $\mathfrak{C}$ by means of a Stieltjes integral and in doing this to extend it from the space of quasi-polynomials to a wider function space.

2. Orthogonal polynomials on a circle

2.1. Assume that the sequence $\{c_k\}_{-\infty}^{\infty}$ is positive relative to the circumference of a circle. This means that all the determinants

$$\Delta_k = \begin{vmatrix} c_0 & c_1 \cdots\cdots & c_k \\ c_{-1} & c_0 \cdots\cdots & c_{k-1} \\ \cdot\cdot\cdot\cdot\cdot\cdot\cdot\cdot\cdot\cdot\cdot \\ \cdot\cdot\cdot\cdot\cdot\cdot\cdot\cdot\cdot\cdot\cdot \\ c_{-k} & c_{-k+1} \cdots & c_0 \end{vmatrix} \quad (k = 0, 1, 2, \ldots)$$

are positive. We also put

$$\Delta_{-1} = 1$$

and introduce the polynomials $P_k(z)$ by the equations

$$P_k(z) = \frac{1}{\Delta_{k-1}} \begin{vmatrix} c_0 & c_1 \cdots\cdots & c_{k-1} & c_k \\ c_{-1} & c_0 \cdots\cdots & c_{k-2} & c_{k-1} \\ \cdot\cdot\cdot\cdot\cdot\cdot\cdot\cdot\cdot\cdot\cdot\cdot\cdot\cdot\cdot\cdot \\ c_{-k+1} & c_{-k+2} \cdots & c_0 & c_1 \\ 1 & z \cdots\cdots & z^{k-1} & z^k \end{vmatrix} = z^k + \ldots$$

$$(k = 0, 1, 2, \ldots).$$

These polynomials possess the following orthogonality properties:

$$\mathfrak{C}\left\{ P_k(z)\frac{1}{z^l} \right\} = 0 \quad (l = 0, 1, 2, \ldots, k-1), \qquad \ldots [5.5a]$$

whence it follows that

$$\mathfrak{C}\left\{ P_k(z)\overline{P_l}\left(\frac{1}{z}\right) \right\} = 0 \quad (k \neq l;\ k, l = 0, 1, 2, \ldots), \qquad \ldots [5.5]$$

where the bar over P signifies that the coefficients in it should be replaced by their complex conjugates. In going over to the Stieltjes integral relation [5.5] takes on the form

$$\frac{1}{2\pi}\int_{-\pi}^{\pi} P_k(e^{i\theta})\overline{P_l}(e^{i\theta})\, d\sigma(\theta) = 0 \quad (k \neq l;\ k, l = 0, 1, 2, \ldots).$$

The polynomials $P_k(z)$ are called *polynomials orthogonal on a circle*. The basic theory of these polynomials was constructed by SZEGÖ[1].

It is not difficult to verify that

$$h_k \equiv \mathfrak{C}\left\{ P_k(z)\overline{P_k}\left(\frac{1}{z}\right) \right\} = \mathfrak{C}\left\{ P_k(z)\frac{1}{z^k} \right\} = \frac{\Delta_k}{\Delta_{k-1}} \quad (k = 0, 1, 2, \ldots).$$

We now state a number of important relations which are mainly formal and which are satisfied by orthogonal polynomials. In doing this we shall use an operation which will be applied to polynomials and denoted by an asterisk; if $R(z)$ is a polynomial of exact degree k then

$$R^*(z) = z^k \overline{R}\left(\frac{1}{z}\right).$$

Further we introduce the constants

$$a_k = \frac{\Delta_{k-1}}{\Delta_k} \mathfrak{C}\{P_k(z)z\}$$

$$= \frac{1}{\Delta_k} \begin{vmatrix} c_0 & c_1 & \cdots\cdots & c_k \\ c_{-1} & c_0 & \cdots\cdots & c_{k-1} \\ \cdots & \cdots & \cdots\cdots\cdots & \cdots \\ c_{-k+1} & c_{-k+2} & \cdots & c_1 \\ c_1 & c_2 & \cdots\cdots & c_{k+1} \end{vmatrix} \quad (k = 1, 2, 3, \ldots) \quad \ldots[5.6]$$

$$a_0 = \frac{c_1}{c_0},$$

and prove that the following relation is valid:

$$P_{k+1}(z) = zP_k(z) - a_k P_k^*(z) \quad (k = 0, 1, 2, \ldots), \qquad \ldots[5.7]$$

to which may be adjoined the equation

$$P_{k+1}^*(z) = P_k^*(z) - \overline{a}_k z P_k(z) \quad (k = 0, 1, 2, \ldots). \qquad \ldots[5.8]$$

To prove relation [5.7] we note that

$$\frac{P_k(z) - z^k}{z^{k-1}} = \sum_{l=0}^{k-1} B_l^{(k)} \overline{P}_l\left(\frac{1}{z}\right),$$

where

$$B_l^{(k)} \frac{\Delta_l}{\Delta_{l-1}} = \mathfrak{C}\left\{ \frac{P_k(z) - z^k}{z^{k-1}} P_l(z) \right\} = \mathfrak{C}\left\{ P_k(z) \frac{P_l(z)}{z^{k-1}} \right\} - \mathfrak{C}\{P_l(z)z\}.$$

By virtue of [5.5a] the first term on the right hand side is zero while the second, according to [5.6], is

$$-\frac{\Delta_l}{\Delta_{l-1}} a_l.$$

Thus we have

$$B_l^{(k)} = -a_l \quad (l = 0, 1, 2, \ldots, k-1)$$

and therefore

$$P_k(z) = z^k - z^{k-1} \sum_{l=0}^{k-1} a_l \overline{P_l}\left(\frac{1}{z}\right),$$

whence [5.7] follows directly.

The system [5.7], [5.8] of two finite-difference equations of first order appears here in place of the single finite-difference equation of second order which we encountered in the case of orthogonal polynomials on a straight line.

With the aid of [5.7] it is not difficult to obtain the equation

$$h_{k+1} = h_k(1 - |a_k|^2) \quad (h_0 = c_0; \ k = 0, 1, 2, \ldots),$$

from which it follows, incidentally, that

$$h_n = \frac{\Delta_n}{\Delta_{n-1}} = c_0 \prod_{k=0}^{n-1} (1 - |a_k|^2).$$

Therefore the inequalities

$$\Delta_k > 0 \quad (k = 0, 1, 2, \ldots),$$

which express the positive character of the sequence $\{c_k\}_{-\infty}^{\infty}$ with respect to the circle, are equivalent to the inequalities

$$|a_k| < 1 \quad (k = 0, 1, 2, \ldots).$$

Among other essentially formal relations we note the *analogue of the Darboux-Christoffel equation*

$$\sum_0^n \frac{P_k(z)\overline{P_k(\zeta)}}{h_k} = \frac{P_n^*(z)\overline{P_n^*(\zeta)} - z\overline{\zeta}P_n(z)\overline{P_n(\zeta)}}{h_n(1 - z\overline{\zeta})}, \qquad \ldots[5.9]$$

the proof of which is left to the reader.

There is an important consequence of this equation.

THEOREM **5.2.1** (CONCERNING ZEROS). *The polynomial $P_n(z)$ has no zeros in the region $|z| \geqq 1$ and therefore the polynomial $P_n^*(z)$ has no zeros in the region $|z| \leqq 1$.*

Proof. We assume to start with that $P_n(z)$ has a zero whose modulus is $|z_0| > 1$. In this case, if we put $z = \zeta = z_0$ in equation [5.9] we obtain

$$\sum_0^n \frac{|P_k(z_0)|^2}{h_k} = \frac{|P_n^*(z_0)|^2}{h_n(1 - |z_0|^2)} \leqq 0,$$

which is absurd. Now we assume that $P_n(z)$ has a zero z_0 whose modulus is $|z_0| = 1$. Then $P_n^*(z)$ will have the same zero. But in this case we have, for any ζ with $|\zeta| < 1$:

$$\sum_0^n \frac{P_k(z_0)\overline{P_k(\zeta)}}{h_k} = 0.$$

This equation remains true also for $\zeta = z_0$, i.e. we have

$$\sum_0^n \frac{|P_k(z_0)|^2}{h_k} = 0,$$

which is absurd.

2.2. As we saw, the trigonometric moment problem is always determinate. However, there exists here also a curious alternative, which we shall now discuss.

Assume that $\sigma(\theta)$ $(-\pi \leqq \theta \leqq \pi)$ is the solution of a trigonometric moment problem. We introduce the space L_σ^2 and construct in the same way as was done in Section 2 of Chapter 2 an operator U which is defined in the space l^2 of sequences

$$x = \{x_0, x_0, x_1, \ldots\} \quad \left(\sum_0^\infty |x_k|^2 < \infty\right),$$

by means of the equation

$$Ux = \underset{\sigma(\theta)}{\text{l.i.m.}} \frac{1}{\sqrt{2\pi}} \sum_0^n x_k \frac{P_k(e^{i\theta})}{\sqrt{h_k}}.$$

This operator is isometric, but not always unitary, i,e. its range need not coincide with the whole space L_σ^2. For instance, in the simplest case, when $\sigma(\theta) = 0$ $(-\pi \leqq \theta \leqq \pi)$ we have $P_k(z) = z^k$ and the functions $F(\theta) = Ux$ prove to be the limiting values on the unit circle of functions which are analytic within that circle. Therefore we have in this case that the manifold on to which the operator U maps the space l^2 certainly does not coincide with L_σ^2.

The question when the operator U is unitary reduces to the following: *under what conditions do the functions $e^{ik\theta}$ $(k = 0, 1, 2, \ldots)$ form a closed system in L_σ^2 i.e. under what conditions do the linear combinations of these functions form a dense set in L_σ^2?* The answer to this question was given by KOLMOGOROV[1, 2] and KREIN[4] and somewhat later the present author showed† that the criterion obtained is valid for L_σ^p with any $p \geqq 1$. This criterion consists of the following: *In order that the*

† See AKHIEZER[2].

system of functions $e^{ik\theta}$ ($k = 0, 1, 2, \dots$) be closed in L_σ^p it is necessary and sufficient that

$$\int_{-\pi}^{\pi} \ln \sigma'(\theta)\, d\theta = -\infty,$$

where $\sigma'(\theta)$ is the derivative of the absolutely continuous part of the function $\sigma(\theta)$.

We shall say that the case C holds if

$$\int_{-\pi}^{\pi} \ln \sigma'(\theta)\, d\theta > -\infty,$$

and the case D if

$$\int_{-\pi}^{\pi} \ln \sigma'(\theta)\, d\theta = -\infty.$$

Thus the operator U is unitary if and only if the case D holds.

THEOREM 5.2.2.† *In order that case D be realized it is necessary that the series*

$$\sum_{0}^{\infty} \frac{|P_k(z)|^2}{h_k} \qquad \dots [5.10]$$

should be divergent at every point of the circle $|z| < 1$ and it is sufficient that it should diverge in the single point $z = 0$.

Proof. This proof is based on the following simple fact, which is analogous to Theorem 2.5.1 and the verification of which we leave to the reader:

Let $H_{n+1}(z)$ range over the aggregate of all polynomials of degree $n+1$ that satisfy the condition

$$H_{n+1}(z_0) = 1,$$

where z_0 is a given point of the circle $|z| < 1$; in this case

$$\min \frac{1}{2\pi} \int_{-\pi}^{\pi} |H_{n+1}(e^{i\theta})|^2\, d\sigma(\theta) = \frac{1}{\displaystyle\sum_{0}^{n+1} \frac{|P_k(z_0)|^2}{h_k}}$$

and this minimum is attained only for the polynomial

$$H_{n+1}^{\text{extr}}(z) = \frac{\displaystyle\sum_{0}^{n+1} \frac{P_k(z)\overline{P_k(z_0)}}{h_k}}{\displaystyle\sum_{0}^{n+1} \frac{|P_k(z_0)|^2}{h_k}}.$$

† See KREIN[4] and GERONIMUS[1] (see also the footnote to Theorem 5.2.3).

Assuming this fact to be established, we take it that the series [5.10] converges at some point z_0 in the circle $|z| < 1$. If this is so we can write down the following inequality for any polynomial $G_n(z)$ of degree n:

$$\frac{1}{2\pi}\int_{-\pi}^{\pi}\left|\frac{1}{z-z_0} - G_n(z)\right|^2 d\sigma(\theta)$$

$$= \frac{1}{2\pi}\int_{-\pi}^{\pi}\frac{1}{|z-z_0|^2}\,|1-(z-z_0)G_n(z)|^2\,d\sigma(\theta)$$

$$\geqq \frac{1}{(1+|z_0|)^2}\frac{1}{2\pi}\int_{-\pi}^{\pi}|H_{n+1}(z)|^2\,d\sigma(\theta)$$

$$\geqq \frac{1}{(1+|z_0|)^2}\frac{1}{\sum\limits_{0}^{n+1}\dfrac{|P_k(z_0)|^2}{h_k}},$$

where everywhere under the integral sign we have $z = e^{i\theta}$. But by virtue of the fact that the series [5.10] is convergent at the point z_0 the right-hand side is greater than a certain positive quantity, for all n. Therefore the function $1/(e^{i\theta} - z_0)$ cannot be approximated with arbitrary accuracy in L_σ^2 by means of any polynomial in $e^{i\theta}$ and therefore the case C must hold.

We assume now that the series [5.10] diverges at the point $z = 0$. In this case the expression

$$\min_{G_n}\frac{1}{2\pi}\int_{-\pi}^{\pi}|e^{-i\theta} - G_n(e^{i\theta})|^2\,d\sigma(\theta)$$

$$= \min_{G_n}\frac{1}{2\pi}\int_{-\pi}^{\pi}|1-zG_n(z)|^2\,d\sigma(\theta) = \frac{1}{\sum\limits_{0}^{n+1}\dfrac{|P_k(0)|^2}{h_k}}$$

tends to zero for $n \to \infty$. This means that the function $e^{-i\theta}$ can be arbitrarily accurately approximated by means of polynomials in $e^{i\theta}$. By use of induction this statement can be extended to all further natural powers of the function $e^{-i\theta}$. The first step in this induction is based on the equation

$$\min_{A,G_n}\frac{1}{2\pi}\int_{-\pi}^{\pi}|e^{-2i\theta} - Ae^{-i\theta} - G_n(e^{i\theta})|^2\,d\sigma(\theta)$$

$$= \min_{A,G_n}\frac{1}{2\pi}\int_{-\pi}^{\pi}|e^{-i\theta} - A - e^{i\theta}G_n(e^{i\theta})|^2\,d\sigma(\theta) = \frac{1}{\sum\limits_{0}^{n+2}\dfrac{|P_k(0)|^2}{h_k}}.$$

The subsequent steps need not be further explained.

But if $e^{-im\theta}$ for $m = 1, 2, 3, \ldots$ can be arbitrarily accurately approximated in L_σ^2 by means of a polynomial in $e^{i\theta}$ then the sequence $\{e^{ik\theta}\}_{-\infty}^{\infty}$ remains closed in L_σ^2 after removal of all functions $e^{-im\theta}$ ($m = 1, 2, 3, \ldots$) and this means that case D holds.

Thus the theorem is proved.

COROLLARY 5.2 (GERONIMUS[1]).   *Case C is realized if and only if*

$$\sum_0^\infty |a_k^2| < \infty.$$

Proof. We take the Darboux-Christoffel formula and in it put $z = \zeta = 0$. Since $P_k^*(0) = 1$, we obtain the equation

$$\sum_0^n \frac{|P_k(0)|^2}{h_k} = \frac{1}{h_n} = \frac{1}{c_0 \prod_{k=0}^{n-1} (1 - |a_k|^2)}.$$

Owing to this equation and to Theorem 5.2.2 we can say that case C is valid if and only if the infinite product

$$\prod_{k=0}^\infty (1 - |a_k|^2)$$

is convergent, and this is just another formulation of the statement to be proved.

2.3. THEOREM 5.2.3.†   *If case C holds the series*

$$\sum_0^\infty \frac{|P_k(z)|^2}{h_k} \qquad \ldots[5.10]$$

and the sequence

$$\{P_n^*(z)\}_0^\infty \qquad \ldots[5.11]$$

converge uniformly in every circle $|z| \leq r < 1$.

Proof. We write down the Darboux-Christoffel formula for $\zeta = z$:

$$\sum_0^n \frac{|P_k(z)|^2}{h_k} + \frac{|z|^2}{1 - |z|^2} \frac{|P_n(z)|^2}{h_n} = \frac{1}{1 - |z|^2} \frac{|P_n^*(z)|^2}{h_n}. \qquad \ldots[5.12]$$

For any z in the circle $|z| < 1$ the left hand side of this formula is not less than

$$\frac{|P_0(z)|^2}{h_0} = \frac{1}{c_0}.$$

---

† This theorem was first proved by SZEGÖ[1] for the case when $\sigma(\theta)$ is absolutely continuous. For the general case, see GERONIMUS[1, 4].

Therefore it follows from $[5.12]$ that

$$\frac{|P_n^*(z)|}{\sqrt{h_n}} \geqq \frac{\sqrt{1-|z|^2}}{\sqrt{c_0}} \quad (|z|<1),$$

whence

$$\left|\frac{1}{P_n^*(z)}\right| \leqq \frac{\sqrt{c_0}}{\sqrt{h_n}\sqrt{1-|z|^2}} \quad (|z|<1). \qquad \ldots[5.13]$$

Further we find from $[5.12]$ for $z=0$ that

$$\frac{1}{h_n} = \sum_0^n \frac{|P_k(0)|^2}{h_k}.$$

Therefore it follows from Theorem **5.2.2** that the quantity $1/h_n$ tends monotonically to a finite limit as $n \to \infty$:

$$\lim_{n\to\infty} \frac{1}{h_n} = \sum_0^\infty \frac{|P_k(0)|^2}{h_k} = \frac{1}{h}.$$

Inequality $[5.13]$ shows that the sequence

$$\left\{\frac{1}{P_n^*(z)}\right\}_0^\infty \qquad \ldots[5.14]$$

is unformly bounded in every circle $|z| \leqq r < 1$. By virtue of the Stieltjes-Vitali theorem one can therefore single out from the sequence $[5.14]$ a sub-sequence

$$\left\{\frac{1}{P_{n_i}^*(z)}\right\}_{i=1}^\infty,$$

which converges uniformly in every circle $|z| \leqq r < 1$. The limit function of this sequence does not become zero in any point of the circle $|z| < 1$. Indeed, it cannot be identically zero, since all functions in the sequence $[5.14]$ are equal to 1 at the point $z=0$. On the other hand, none of the functions of the sequence become zero in the circle $|z| \leqq 1$ and therefore it follows by Hurwitz's theorem that the limit function also cannot become zero within this region. Hence the sequence

$$\{P_{n_i}^*(z)\}_{i=1}^\infty \qquad \ldots[5.11a]$$

tends uniformly to a limit in every circle $|z| \leqq r < 1$. Therefore it follows from $[5.12]$ that the sum of the series $[5.10]$ is continuous in the circle $|z| < 1$. And since the terms of this series are $\geqq 0$ the series

$[5.10]$ must converge uniformly, by Dini's theorem, in any interior circle $|z| \leqq r < 1$.

Thus the first statement of the theorem is proved.

From it, it follows that in every circle $|z| \leqq r < 1$ the series

$$\sum_0^\infty \frac{P_k(z)\overline{P_k(0)}}{h_k}$$

converges uniformly. It is then easy to conclude that not only the sequence $[5.11a]$ but also the whole sequence $[5.11]$ converges uniformly in every circle $|z| \leqq r < 1$ and this proves the second statement of the theorem. One merely has to take the Darboux-Christoffel theorem for $\zeta = 0$, which has the form

$$P_n^*(z) = h_n \sum_0^n \frac{P_k(z)\overline{P_k(0)}}{h_k}.$$

COROLLARY 5.2.3. *If case C holds there exists a function $\Delta(z)$ analytic in the region $|z| > 1$ which does not become zero anywhere in this region, such that*

$$\lim_{n \to \infty} \frac{P_n(z)}{z^n \Delta(z)} = 1$$

at every point z of the region $|z| > 1$ and the convergence is uniform in every circle $|z| \geqq R > 1$.

Indeed this function $\Delta(z)$ is

$$\lim_{n \to \infty} \overline{P_n^*}\!\left(\frac{1}{z}\right) \quad (|z| > 1).$$

The existence of this limit and the properties just formulated for it follow from Theorem 5.2.3.

3. Hermitian-positive functions of a single argument

3.1. The continuous analogue of a sequence $\{c_k\}_{-\infty}^\infty$ which is positive with respect to a circular boundary is a function $f(x)$ $(-\infty < x < \infty)$, such that for any positive integer n, any $x_k \geqq 0$ $(k = 1, 2, ..., n)$ and any complex ξ_k $(k = 1, 2, ..., n)$ the inequality

$$\sum_{\alpha, \beta = 1}^n f(x_\alpha - x_\beta)\xi_\alpha \bar{\xi}_\beta \geqq 0. \qquad \qquad ...[5.15]$$

is valid. Functions which satisfy this condition are called *hermitian-positive*.

As it is customary to say in the theory of integral equations, the inequalities [5.15] express the fact that the kernel

$$K(x, y) = f(x-y) \quad (0 \leq x, y < \infty)$$

is hermitian-positive. One can introduce the class of functions $f(x)$ for which this kernel is hermitian-positive only for $0 \leq x, y < a$. It is clear that in this case the function $f(x)$ has to be given only in the interval $(-a, a)$. Such functions can be considered as continuous analogues of finite sequences $\{c_k\}_{-n}^{n}$ for which

$$\sum_{\alpha, \beta=0}^{n} c_{\alpha-\beta} \xi_\alpha \bar{\xi}_\beta \geq 0.$$

The study of such sequences has a special algebraic interest, but is beyond the scope of this book. As for hermitian-positive functions in a finite interval, we shall not dwell on the theory relating to them.

DEFINITION 5.3.1. We shall say that the function $f(x)$ $(-a < x < a)$ *belongs to the class* P_a, if it is hermitian-positive in the interval $(-a, a)$ and is continuous at the point $x = 0$.

Let us clarify what consequences follow for the function $f(x)$ from its hermitian-positive property, i.e. the property expressed by inequality [5.15]. To do this we write down this inequality in two cases: first for $n = 2$, $x_1 = x$, $x_2 = 0$, and then for $n = 3$, $x_3 = 0$, $\xi_1 = \xi$, $\xi_2 = -\xi$, $\xi_3 = \eta$. The first inequality takes on the form

$$f(0) \mid \xi_1 \mid^2 + f(x)\xi_1\bar{\xi}_2 + f(-x)\bar{\xi}_1\xi_2 + f(0) \mid \xi_2 \mid^2 \geq 0,$$

whence it follows immediately that

$$f(0) \geq 0, \quad f(-x) = \overline{f(x)}, \quad \mid f(x) \mid \leq f(0).$$

The second inequality appears in the form

$$f(0) \mid \eta \mid^2 + 2 \operatorname{Re} \{[f(x_1) - f(x_2)]\xi\bar{\eta}\}$$
$$+ 2\{f(0) - \operatorname{Re} f(x_1 - x_2)\} \mid \xi \mid^2 \geq 0,$$

whence it follows that

$$\mid f(x_1) - f(x_2) \mid^2 \leq 2f(0)\{f(0) - \operatorname{Re} f(x_1 - x_2)\}. \quad \ldots[5.16]$$

Thus it follows from the hermitian-positive property of the function $f(x)$ alone that this function is bounded and also that if this function is to be uniformly continuous in the whole interval $(-a, a)$ it is necessary and sufficient that its real part be continuous at only the one point† $x = 0$. In particular a function $f(x)$ of the class P_a is uniformly continuous and if it is not identically zero then $f(0) > 0$. Therefore one could put $f(0) = 1$ without restricting generality.

3.2. The basic property of hermitian-positive functions is that they allow a special integral representation. For $a = \infty$ this representation was derived in full generality‡ by BOCHNER[1] and for $a < \infty$ by KREIN[1]. From this integral representation it follows that with the normalization $f(0) = 1$ the class P_∞ coincides with the aggregate of characteristic functions for all probability distribution laws.

THEOREM **5.3.2.** *In order that a function* $f(x)$ $(-a < x < a)$ *may have a representation of the form*

$$f(x) = \int_{-\infty}^{\infty} e^{ixt} \, d\sigma(t), \qquad \dots[5.17]$$

where $\sigma(t)$ $(-\infty \leq t \leq \infty)$ *is a non-decreasing function of bounded variation it is necessary and sufficient that* $f(x) \in P_a$.

Proof. The fact that the condition is necessary can be verified directly. Indeed, it follows from the representation $[5.17]$ that

$$\sum_{\alpha, \beta = 1}^{n} f(x_\alpha - x_\beta)\xi_\alpha\bar{\xi}_\beta = \int_{-\infty}^{\infty} \left| \sum_{1}^{n} \xi_k e^{itx_k} \right|^2 d\sigma(t) \geqq 0,$$

and the continuity of the function $f(x)$ is obvious.

The proof of sufficiency is somewhat complicated.

If the representation $[5.17]$ is proved for $-a < x < a$, then the function $f(x)$ is shown to be continued on the whole axis $-\infty < x < \infty$, maintaining its hermitian-positive property. Conversely, if by some means one proves the possibility of such a continuation, then the representation $[5.17]$ can be obtained by applying Bochner's theorem for the case $a = \infty$. In the original proof due to Krein just this possibility of continuation was established, and Bochner's theorem then used.

Here we give a proof of sufficiency due to LIVSHITS[1] which is applicable equally for both $a < \infty$ and $a = \infty$. This proof is based on the general Theorem **4.3.3.**

For this purpose we take for G the aggregate of all functions of the form

$$\phi(s) = \int_{\alpha}^{\beta} e^{isx} \, d\mu(x),$$

---

† This last fact was first discovered by Artemenko. According to a verbal communication by Krein the first proof by Artemenko was rather complicated. The inequality $[5.16]$ and its proof as given here are due to Krein.

‡ With some restrictions this representation was obtained by Khinchin. See also MATHIAS[1].

where $[\alpha, \beta]$ is any interval included in $(-a, a)$, and $\mu(x)$ is any, in general complex, function of bounded variation. We then construct $\mathfrak{M}$ and define a functional $\mathfrak{P}$ starting from the equation

$$\mathfrak{P}\{e^{ix_1 s} e^{-ix_2 s}\} = f(x_1 - x_2).$$

Now we must only verify that the conditions of Theorem **4.3.3** are satisfied. Condition a) is evidently satisfied. As the set D we take the aggregate of all functions of the form

$$\psi(s) = \int_\alpha^\beta e^{isx} \gamma(x)\, dx,$$

where $\gamma(x)$ is continuous. We have first of all to convince ourselves that this set D is dense in G with the norm

$$\| \phi \| = \sqrt{\mathfrak{P}\{\phi(s)\overline{\phi(s)}\}}.$$

But to do this it is sufficient to prove that the function $\phi_0(s) = e^{ics}$ for any $c \in (-a, a)$ may be arbitrarily closely approximated by a function $\psi(s) \in D$. This fact may be proved very simply. Indeed we take

$$\psi_0(s) = \frac{1}{c_2 - c_1} \int_{c_1}^{c_2} e^{isx}\, dx \in D,$$

where $c \in (c_1, c_2)$ and $c_2 - c_1$ is arbitrarily small. We then obtain the result that

$$\mathfrak{P}\{[\phi_0(s) - \psi_0(s)][\overline{\phi_0(s) - \psi_0(s)}]\}$$

$$= \frac{1}{(c_2 - c_1)^2} \int_{c_1}^{c_2} \int_{c_1}^{c_2} \mathfrak{P}\{(e^{ics} - e^{isx})(e^{-ics} - e^{-isy})\}\, dx\, dy$$

$$= \frac{1}{(c_2 - c_1)^2} \int_{c_1}^{c_2} \int_{c_1}^{c_2} \{f(0) - f(c - y) - f(x - c) + f(x - y)\}\, dx\, dy,$$

and it remains to note the fact that $f(x)$ is continuous. Then we must verify the fact that condition b) is satisfied and it is sufficient also to do this for the function e^{ics} for any $c \in (-a, a)$. This is done as follows:

$$\frac{e^{ics} - e^{ict}}{s - t} = i \int_0^c e^{ixs} e^{it(c-x)}\, dx \in G.$$

It is worth noting that the function $\sigma(t)$ in the integral representation [5.17] is certainly essentially unique if $a = \infty$. A simple proof of this fact for a more general case is given in the following section.

4. Hermitian-positive functions in many-dimensional spaces

4.1. DEFINITION 5.4.1.   A function $f(x) = f(x_1, x_2, ..., x_m)$ $(-\infty < x_1, x_2, ..., x_m < \infty)$ *belongs to the class* $P(E_m)$, if it is continuous at the point 0 and is hermitian-positive, i.e. if for any positive integral n and for any vectors x^k $(k = 1, 2, ..., n)$ the inequality

$$\sum_{\alpha, \beta = 1}^{n} f(x^\alpha - x^\beta)\xi_\alpha \bar{\xi}_\beta \geqq 0 \qquad ...[5.18]$$

holds whatever we choose for the complex numbers ξ_k $(k = 1, 2, ..., n)$.

Just as in the one-dimensional case, one proves that the function $f(x) \in P(E_m)$ is uniformly continuous in the whole space and satisfies the relations

$$f(-x) = \overline{f(x)}, \quad |f(x)| \leqq f(0).$$

As in the one-dimensional case the central place in the theory is occupied by the theorem on the integral representation.   In this theorem a bounded† measure occurs i.e. a bounded non-negative fully additive function over the family of all Borel sets $A \in E_m$. We denote by A_t the set of all points $x \in E_m$ for which $x_j \leqq t_j$ $(j = 1, 2, ..., m)$ and we put

$$\omega(t) = \omega(A_t).$$

$\omega(t)$ is a non-decreasing bounded function of the point $t \in E_m$. Sometimes we shall say that it is generated by the measure $\omega(A)$. We also introduce intervals i.e. aggregates of all points x for which $a_j < x_j < b_j$ $(j = 1, 2, ..., m)$, and denote them by $I_{a, b} = I$. An interval I is called an interval of continuity if

$$\omega(I) = \omega(\bar{I}),$$

where $\bar{I}$ is the closure of I.   One can always find a set Q of real numbers which is at most denumerable and such that if one of the numbers a_j, b_j $(j = 1, 2, ..., m)$ does not belong to Q then the interval $I_{a, b}$ is an interval of continuity.   Two measures $\omega_1(A)$ and $\omega_2(A)$ are equal if for every finite continuous function $\phi(t)$ the equality

$$\int_{E_m} \phi(t) \, d\omega_1(t) = \int_{E_m} \phi(t) \, d\omega_2(t)$$

holds.   For two measures to be equal it is necessary and sufficient that their difference should become zero in every interval in which it is continuous.   For a sequence of uniformly bounded measures $\{\omega_n(A)\}_1^\infty$ a

† In the following we shall have to deal only with bounded measures and can therefore simply call them measures.

theorem holds which is analogous to the first Helly theorem (it is due to Radon). By this theorem there exists a sub-sequence $\{\omega_{n_k}(A)\}_{k=1}^{\infty}$ and a measure $\omega(A)$ such that on any one of its intervals of continuity I:

$$\lim_{k \to \infty} \omega_{n_k}(I) = \omega(I).$$

The analogue of HELLY'S second theorem† is also valid.

THEOREM **5.4.1** (BOCHNER[2, 3]). *Any function $f(x) \in P(E_m)$ permits the representation*

$$f(x) = \int_{E_m} e^{i(x, t)} \, d\omega(t), \qquad \ldots[5.19]$$

where $(x, t) = x_1 t_1 + x_2 t_2 + \ldots + x_m t_m$, and $\omega(t)$ is a monotonic point function generated by a certain measure $\omega(A)$ $(A \subset E_m)$. The measure is uniquely determined by the function $f(x)$.

Conversely any function $f(x)$ which has a representation $[5.19]$ belongs to the class $P(E_m)$.

Proof. The last statement can be accepted as trivial.

We turn to the first statement and in the first place note that by virtue of the inequalities $[5.18]$ the integral inequality

$$\int_{E_m} \int_{E_m} f(x-y)\rho(x)\rho(y) \, dv(x) \, dv(y) \geqq 0, \qquad \ldots[5.20]$$

is valid for any function $\rho(x) \in L^1(E_m)$ where

$$dv(x) = dx_1 \, dx_2 \ldots dx_m, \quad dv(y) = dy_1 \, dy_2 \ldots dy_m.$$

We put

$$\rho(x) = e^{-2\varepsilon|x|^2 - i(x, t)},$$

where $\varepsilon > 0$ and t are fixed and $|x|^2$ is the square of the length of the vector x. We note that

$$2\{|x|^2 + |y|^2\} = |x-y|^2 + |x+y|^2,$$

and in the integral we make the substitution

$$x - y = z, \quad x + y = u.$$

---

† The general theory of completely additive functions of sets in m-dimensional space does not seem to be given in any text book. We must therefore refer the reader for the facts enumerated above to the original papers by RADON[1] and BOCHNER[2] (see also HAVILAND[1]).

Then the inequality [5.20] takes on the form

$$\int_{E_m} e^{-\varepsilon|u|^2}\, dv(u) \int_{E_m} e^{-\varepsilon|z|^2} f(z) e^{-i(z,\,t)}\, dv(z) \geqq 0.$$

This means that the Fourier transform

$$\phi_\varepsilon(t) = \frac{1}{(\sqrt{2\pi})^m} \int_{E_m} f(z) e^{-\varepsilon|z|^2} e^{-i(z,\,t)}\, dv(z) \qquad \dots[5.21]$$

of the function $f(z) e^{-\varepsilon|z|^2} \in L^2(E_m)$ is everywhere greater than or equal to zero, for any $\varepsilon > 0$.

We prove that $\phi_\varepsilon(t) \in L^1(E_m)$. To do this we write down the Parseval equality

$$\int_{E_m} \phi_\varepsilon(t) e^{-\frac{1}{2}\delta|t|^2}\, dv(t) = \int_{E_m} f(z) e^{-\varepsilon|z|^2} \frac{1}{\sqrt{\delta^m}} e^{-\frac{|z|^2}{2\delta}}\, dv(z).$$

Hence for any $\delta > 0$ we have

$$\int_{E_m} \phi_\varepsilon(t) e^{-\frac{1}{2}\delta|t|^2}\, dv(t) \leqq f(0) \frac{1}{\sqrt{\delta^m}} \int_{E_m} e^{-\frac{|z|^2}{2\delta}}\, dv(z) = (\sqrt{2\pi})^m f(0).$$

Therefore

$$\int_{E_m} \phi_\varepsilon(t)\, dv(t) \leqq (\sqrt{2\pi})^m f(0),$$

i.e. $\phi_\varepsilon(t)$ does indeed belong to $L^1(E_m)$.

Hence from a well-known theorem in the theory of Fourier integrals it follows that formula [5.21] can be inverted:

$$f(z) e^{-\varepsilon|z|^2} = \frac{1}{(\sqrt{2\pi})^m} \int_{E_m} \phi_\varepsilon(t) e^{i(t,\,z)}\, dv(t)$$

or

$$f(z) e^{-\varepsilon|z|^2} = \int_{E_m} e^{i(t,\,z)}\, d\omega_\varepsilon(t), \qquad\qquad \dots[5.22]$$

where

$$\omega_\varepsilon(A) = \frac{1}{(\sqrt{2\pi})^m} \int_A \phi_\varepsilon(t)\, dv(t) \quad (A \subset E_m).$$

Since

$$\omega_\varepsilon(E_m) = f(0),$$

one can apply Helly's first theorem to the aggregate of measures $\omega_a(A)$ $(\varepsilon > 0)$ and therefore one can find a sequence $\{\omega_{\varepsilon_k}(A)\}_{k=1}^\infty$ and a limiting measure $\omega(A)$.

Now we must apply Helly's second theorem to equation $[5.22]$. To do this we have to convince ourselves that the quantity

$$\int_{|t|>N} d\omega_\varepsilon(t)$$

will be arbitrarily small for all $\varepsilon > 0$, provided only N is sufficiently large. To do this we integrate $[5.22]$ over each of the variables z_i from $-\gamma$ to γ. We find that

$$f(0) - \frac{1}{(2\gamma)^m}\int_{-\gamma}^{\gamma} \cdots \int_{-\gamma}^{\gamma} f(z)e^{-\varepsilon|z|^2}\, dz_1 \ldots dz_m$$

$$= \int_{E_m}\left\{1 - \prod_1^m \frac{\sin \gamma t_k}{\gamma t_k}\right\}d\omega_\varepsilon(t).$$

Taking $N > 2/\gamma$, we thus obtain the inequality

$$\int_{|t|>N} d\omega_\varepsilon(t) \leqq 2\left\{f(0) - \frac{1}{(2\gamma)^m}\int_{-\gamma}^{\gamma} \cdots \int_{-\gamma}^{\gamma} f(z)e^{-\varepsilon|z|^2}\, dz_1 \ldots dz_m\right\},$$

the right hand side of which will be arbitrarily small, uniformly for $\varepsilon \in [0, 1]$ provided only γ is sufficiently small. Thus Helly's second theorem is applicable to equation $[5.22]$. As a result we obtain just the representation $[5.19]$.

The proof of the uniqueness of this representation is very simple. Indeed, we take any interval of continuity $I_{a, b}$ occurring in the representation $[5.19]$ for the measure $\omega(A)$ and multiply both sides of the equation $[5.19]$ by

$$\frac{1}{(2\pi)^m}\prod_1^m \frac{e^{-ib_kx_k} - e^{-ia_kx_k}}{-ix_k} = \psi(x).$$

We then integrate over each of the variables x_k from $-c$ to c. Going to the limit $c \to \infty$, in analogy with the procedure in the one-dimensional case,† we obtain the equation

$$\lim_{c\to\infty}\int_{-c}^{c}\int_{-c}^{c} \cdots \int_{-c}^{c} f(x)\psi(x)\, dx_1\, dx_2 \ldots dx_m = \int_{I_{a, b}} d\omega(t) = \omega(I_{a, b}).$$

Thus the measure of any interval of continuity is determined uniquely by the function $f(x)$ and hence it follows that the representation $[5.19]$ is unique.

† See e.g. GNEDENKO[2].

4.2. DEFINITION **5.4.2.** The function $F(r)$ $(r \geqq 0)$ *belongs to the class* $R(E_m)$ *and is called a radial hermitian-positive function in m-dimensional space if*

$$f(x) = F(\sqrt{x_1^2 + x_2^2 + \ldots + x_m^2}) \in P(E_m).$$

Thus $R(E_m)$ is the aggregate of those functions in $P(E_m)$ which in fact depend only on the distance of the point in m-dimensional space from the origin. It is clear that any such function is real.

It is not difficult to see that

$$R(E_1) \supset R(E_2) \supset R(E_3) \supset \ldots .$$

THEOREM **5.4.2** (SCHOENBERG[1]). *The class* $R(E_m)$ *coincides with the aggregate of all functions of the form*

$$F(r) = \int_0^\infty \Omega_m(ru) \, d\sigma(u), \qquad \ldots [5.23]$$

where $\sigma(u)$ $(u \geqq 0)$ *is a non-decreasing bounded function and* $\Omega_m(z)$ *is defined by the equation*

$$\Omega_m(z) = \Gamma\left(\frac{m}{2}\right)\left(\frac{2}{z}\right)^{\frac{1}{2}(m-2)} J_{\frac{1}{2}(m-2)}(z)$$

$$= 1 - \frac{z^2}{2m} + \frac{z^4}{2 \cdot 4 \cdot m(m+2)} - \frac{z^6}{2 \cdot 4 \cdot 6 \cdot m(m+2)(m+4)} + \ldots .$$

Proof. If $F(r) \in R(E_m)$, then in any case the representation

$$F(r) = \int_{E_m} e^{i(x,\,t)} \, d\omega(t) \qquad \ldots [5.24]$$

holds where $r = |x|$. We assume to start with that $m = 1$. In this case the representation $[5.24]$ can be rewritten in the form

$$F(r) = \int_{-\infty}^\infty e^{\pm irt} \, d\omega(t)$$

Hence

$$F(r) = \int_{-\infty}^\infty \cos(rt) \, d\omega(t) = \int_0^\infty \cos(rt) \, d\sigma(t), \qquad \ldots [5.25]$$

where

$$\sigma(t) = \omega(t) - \omega(-t) \quad (t \geqq 0).$$

But $[5.25]$ is just equation $[5.23]$ for $m = 1$ since $\Omega_1(z) = \cos z$. Now we assume that $m \geqq 2$. We take a fixed $r > 0$ and introduce a unit vector $\xi = x/r$. Further, we denote by S_m the unit sphere in m-dimensional space and also its surface area. If one uses spherical polar co-ordinates in m-dimensional space one finds for the area of an elementary zone on the surface S_m the easily derived expression

$$dS_m(\xi) = 2\frac{(\sqrt{\pi})^{m-1}}{\Gamma\left(\dfrac{m-1}{2}\right)} \sin^{m-2} \theta \, d\theta.$$

We average relation [5.24] over the surface S_m. This gives us the equation

$$F(r) = \int_{E_m} \left\{ \frac{1}{S_m} \int_{S_m} e^{ir(\xi,\, t)} \, dS_m(\xi) \right\} d\omega(t). \qquad \ldots [5.26]$$

We now note that with a suitable choice for the direction of the polar co-ordinate axis we have

$$\int_{S_m} e^{ir(\xi,\, t)} \, dS_m(\xi) = 2\frac{(\sqrt{\pi})^{m-1}}{\Gamma\left(\dfrac{m-1}{2}\right)} \int_0^\pi e^{ir|t|\cos\theta} \sin^{m-2}\theta \, d\theta.$$

On the other hand, we have from the well-known Poisson formula in the theory of Bessel functions (or by direct calculation)

$$\int_0^\pi e^{iz\cos\theta} \sin^{m-2}\theta \, d\theta = \Omega_m(z) \int_0^\pi \sin^{m-2}\theta \, d\theta.$$

Therefore

$$\frac{1}{S_m} \int_{S_m} e^{ir(\xi,\, t)} \, dS_m(\xi) = \Omega_m(r\,|\,t\,|) \qquad \ldots [5.27]$$

and thus

$$F(r) = \int_{E_m} \Omega_m(r\,|\,t\,|) \, d\omega(t).$$

Now it remains to put

$$\sigma(u) = \int_{|t|\leq u} d\omega(t)$$

and equation [5.26] takes on the form [5.23]. Thus it is proved that any function $F(r) \in R(E_m)$ permits the representation [5.23].

To prove the converse statement, it is sufficient to note that by virtue of equation [5.27] the function $\Omega_m(ru)$, considered as a function of r, belongs to $R(E_m)$ for any fixed value $u \geqq 0$.

4.3. In this section we discuss the intersection $R(E_\infty)$ of all classes $R(E_m)$ ($m = 1, 2, 3, ...$). To start with, we prove that this intersection is not empty. To do this we take the equation

$$e^{-\frac{1}{2}r^2} = \frac{1}{(\sqrt{2})^{m-2}\Gamma\left(\dfrac{m}{2}\right)} \int_0^\infty \Omega_m(ru)e^{-\frac{1}{2}u^2}u^{m-1}\,du,$$

which can be obtained by various methods (for instance with the aid of [5.27] or by term-by-term integration of the series). It follows from this equation that for any positive integer m and any fixed $\beta > 0$ we have

$$e^{-\beta r^2} = \int_0^\infty \Omega_m(ru)\,d\alpha_m\left(\frac{u}{\sqrt{2\beta}}\right),$$

where

$$\alpha_m(t) = \frac{1}{\Gamma\left(\dfrac{m}{2}\right)} \int_0^{\frac{1}{2}t^2} e^{-z}z^{\frac{1}{2}m-1}\,dz \quad (0 \leqq t \leqq \infty)$$

is a non-decreasing function of t for which

$$\alpha_m(0) = 0, \quad \alpha_m(\infty) = 1.$$

Thus the function $e^{-\beta r^2}$ for any fixed $\beta > 0$, and also for $\beta = 0$, belongs to the intersection of all classes $R(E_m)$ and therefore this intersection is not empty. The class $R(E_\infty)$ evidently represents the aggregate of all radial hermitian-positive functions in the Hilbert space l^2.

THEOREM **5.4.3** (SCHOENBERG[1]). *In order that the function $F(r)$ ($r \geqq 0$) be representable in the form*

$$F(r) = \int_0^\infty e^{-ur^2}\,d\alpha(u) \quad (r \geqq 0),$$

where $\alpha(u)$ is a non-decreasing bounded function, it is necessary and sufficient that $F(r) \in R(E_\infty)$.

Proof. The necessity of this condition is a trivial consequence of the fact proved above that the function $e^{-\beta r^2}$ belongs to the class $R(E_\infty)$ for any $\beta \geqq 0$.

In order to prove sufficiency we note to start with that the function $\Omega_m(z)$ has two properties that follow from the simplest facts in the theory of Bessel functions:

The function $\Omega_m(z)$ satisfies the differential equation

$$zy'' + (m-1)y' + zy = 0. \qquad [A]$$

The relation

$$\frac{d}{dz}\Omega_m(z) = -\frac{1}{m}z\Omega_{m+2}(z) \qquad [B]$$

holds.

We utilise property [A] to obtain the following inequality

$$|\Omega_m(x)| \leq 1 \quad (x \geq 0;\ m \geq 1). \qquad \ldots[5.28]$$

Indeed we take the function†

$$y(x) = [\Omega_m(x)]^2 + [\Omega'_m(x)]^2.$$

Differentiating this equation with respect to x and taking into account equation [A] we find that

$$y'(x) = 2\Omega'_m(x)\{\Omega_m(x) + \Omega''_m(x)\} = -\frac{2}{x}(m-1)[\Omega'_m(x)]^2 \leq 0.$$

Thus for $m \geq 1$ the function $y(x)$ does not increase. Therefore

$$y(x) \leq y(0) = 1,$$

whence inequality [5.28] follows.

Now we consider the function

$$\psi_m(u, a) = \frac{2}{a^2}\int_0^a \{1 - \Omega_m(ut\sqrt{2m})\}t\,dt,$$

which by virtue of [5.28] satisfies the inequality

$$\psi_m(u, a) \geq 0 \quad (m \geq 1, u \geq 0, a > 0). \qquad \ldots[5.29]$$

For the following this estimate is not sufficient. Putting $m \geq 3$ and $au > 2$ we can use equation [B] and obtain a better estimate. Indeed by virtue of that identity we have

$$\psi_m(u, a) = 1 - \frac{1}{a^2 u^2 m}\int_0^{au\sqrt{2m}} \Omega_m(z)\,z\,dz$$

$$= 1 - \frac{m-2}{a^2 u^2 m}\{1 - \Omega_{m-2}(au\sqrt{2m})\}.$$

Hence using [5.28] we find for $au > 2$ and $m \geq 3$:

$$\psi_m(u, a) \geq \tfrac{1}{2}. \qquad \ldots[5.30]$$

† Here we use a simple but rather general procedure due to Sonin (see SONIN[1] and also SZEGÖ[1]).

Now we can proceed to prove the sufficiency of the condition in Theorem **5.4.3.** Assume that the function $F(r)$ belongs to $R(E_\infty)$. Then it belongs to $R(E_m)$ also for any positive integer m. Therefore we have for any positive integer $m \geqq 3$

$$F(r) = \int_0^\infty \Omega_m(ru) \, d\sigma_m(u) = \int_0^\infty \Omega_m(r\sqrt{2mu}) \, d\alpha_m(u), \quad \dots[5.31]$$

where $\alpha_m(u) = \sigma_m(\sqrt{2mu})$ is a non-decreasing function for which

$$0 \leqq \alpha_m(u) \leqq 1 \quad (u \geqq 0).$$

By Helly's theorem one can find a sequence $\{\alpha_{m_i}(u)\}_{i=1}^\infty$ and a non-decreasing function $\alpha(u)$ such that at all points of continuity of $\alpha(u)$ we have

$$\alpha(u) = \lim_{i \to \infty} \alpha_{m_i}(u).$$

In order to proceed to the limit in equation [5.31] we note in the first place that by virtue of the definition of the function $\Omega_m(z)$ as a series we have

$$\lim_{m \to \infty} \Omega_m(z\sqrt{2m}) = e^{-z^2}$$

uniformly in any finite interval of the real axis. This however is not sufficient because of the infinite extent of the range of integration. In order to remove the difficulty we write down the equation

$$F(0) - F(r) = \int_0^\infty \{1 - \Omega_m(r\sqrt{2mu})\} \, d\alpha_m(u) \quad \dots[5.31a]$$

and, choosing an arbitrarily small number $\varepsilon > 0$ we find a number $a > 0$ such that the inequality

$$0 < \frac{2}{a^2} \int_0^a \{F(0) - F(r)\} r \, dr \leqq \varepsilon$$

holds. Such a choice of a is possible because the function $F(r)$ is continuous.

We then multiply both sides of equation [5.31a] by $2r \, dr/a^2$ and integrate from 0 to a. This gives the equation

$$\frac{2}{a^2} \int_0^a \{F(0) - F(r)\} r \, dr = \int_0^\infty \psi_m(\sqrt{u}, a) \, d\alpha_m(u).$$

Hence, as a result of the choice of a and of the inequalities [5.29] and [5.30] we conclude that

$$\int_N^\infty d\alpha_m(u) \leqq 2 \int_N^\infty \psi_m(\sqrt{u}, a) \, d\alpha_m(u) \leqq 2\varepsilon, \qquad \dots [5.32]$$

provided only $a\sqrt{N} > 2$. We shall in addition assume that N is a point of continuity of $\alpha(u)$.

It is now simple to perform the limiting process in equation [5.31]. Indeed it follows from [5.28] and [5.32] that

$$\left| F(r) - \int_0^N \Omega_m(r\sqrt{2mu}) \, d\alpha_m(u) \right| \leqq 2\varepsilon.$$

Hence by virtue of Helly's theorem we have

$$\left| F(r) - \int_0^N e^{-ur^2} \, d\alpha(u) \right| \leqq 2\varepsilon.$$

Increasing N to infinity we find that

$$F(r) = \int_0^\infty e^{-ur^2} \, d\alpha(u).$$

5. Absolutely monotonic and exponentially convex functions

5.1. The class of functions which will be discussed here was first introduced and studied by BERNSTEIN[1, 2]. A part of the results of Bernstein's second paper was obtained somewhat earlier by HAUSDORFF[1]. Another approach and also some further development is due to WIDDER[1] and to SCHOENBERG[1].

DEFINITION 5.5.1 (BERNSTEIN[1]). A function $f(x)$ $(a < x < b)$ is called *absolutely monotonic* (belonging to the class $\mathfrak{M}(a, b)$) if it satisfies the inequalities

$$f(x) \geqq 0 \quad \text{and} \quad \Delta_h^n f(x) = \sum_{k=0}^n (-1)^{n-k} \binom{n}{k} f(x+kh) \geqq 0$$

for all positive integers n and all x, h for which

$$a < x < b, \quad a < x+nh < b.$$

The trivial case when $f(x) \equiv \infty$ is assumed to be excluded.

LEMMA 5.5.1. *If* $f(x) \in \mathfrak{M}(a, b)$, *then* $f(x)$ *is continuous in the interval* (a, b) *and has a finite limit* $f(a+0)$ *for* $x \to a$.

Proof. Since $f(x)$ is a non-decreasing function by definition, there exists at any point x of the interval (a, b) a limit from the right $f(x+0)$

and a limit from the left $f(x-0)$. In addition $f(a+0)$ and $f(b-0)$ exist. But the function $f(x)$ is $\geqq 0$ everywhere. Therefore the quantity $f(a+0)$ is either finite or equal to $+\infty$. The latter is possible only if $f(x) = \infty$ everywhere, which is excluded. This proves the second statement of the lemma.

In order to prove the first statement we take an arbitrary point $x \in (a, b)$ and a sufficiently small $h > 0$ and write down the inequalities

$$\Delta_h^2 f(x) \geqq 0, \quad \Delta_h^2 f(x-h) \geqq 0.$$

It follows that

$$2f(x+h) \leqq f(x)+f(x+2h), \quad 2f(x) \leqq f(x+h)+f(x-h).$$

Therefore

$$f(x+0) \leqq f(x), \quad 2f(x) \leqq f(x+0)+f(x-0)$$

and thus

$$f(x+0) \leqq f(x-0).$$

Since as a result of the monotonic property we have

$$f(x+0) \geqq f(x-0),$$

it follows that

$$f(x+0) = f(x-0),$$

which expresses the continuity of $f(x)$. Therefore the first statement is also proved.

THEOREM 5.5.1 (BERNSTEIN[2]). *The function $f(x)$ $(-\infty < x \leqq 0)$ can be represented in the form*

$$f(x) = \int_0^\infty e^{xu} \, d\sigma(u), \qquad \qquad ...[5.33]$$

where $\sigma(u)$ is a non-decreasing function of bounded variation, if and only if $f(x)$ is absolutely monotonic in the whole interval $-\infty < x < 0$ and $f(0) = f(-0) < \infty$.

Proof. Assume that the function $f(x)$ has the form $[5.33]$; then we have

$$\Delta_h^n f(x) = \int_0^\infty e^{xu} \sum_{k=0}^n (-1)^{n-k} \binom{n}{k} e^{khu} \, d\sigma(u) = \int_0^\infty e^{xu}(e^{hu}-1)^n \, d\sigma(u),$$

and this expression is $\geqq 0$ for any $n \geqq 0$, $h > 0$ and $x \in (-\infty, 0)$, provided only $x+nh \leqq 0$. Thus the absolute monotonic property of the function $f(x)$ is a necessary condition for it to be representable in the form [5.33]. Similarly the inequality

$$f(0) = \int_0^\infty d\sigma(u) < \infty$$

is necessary.

The proof of sufficiency to which we now turn will be conducted not according to the original method due to Bernstein but in a somewhat simpler fashion, using results pertaining to Hausdorff's moment problem.

Assume that $f(x)$ $(-\infty < x < 0)$ is an absolutely monotonic function and assume that $f(0) = f(-0) < \infty$. We take an arbitrary number $h > 0$ and consider the Hausdorff moment problem

$$f(-kh) = \int_0^1 t^k \, d\omega(t) \quad (k = 0, 1, 2, ...). \qquad ...[5.34]$$

Since for any integers $n \geq 0$ and $m \geq 0$ the moments $s_k = f(-kh)$ satisfy the inequality

$$\sum_{k=0}^n (-1)^k \binom{n}{k} s_{m+k} = \sum_{k=0}^n (-1)^{n-k} \binom{n}{k} f(-mh-nh+kh)$$

$$= \Delta_h^n f(-mh-nh) \geq 0,$$

we have by Theorem 2.6.4 that the moment problem [5.34] has a non-decreasing solution $\omega(u) = \omega_h(u)$.

We make the substitution of variables $t = e^{-hu}$ and rewrite equation [5.34] in the form

$$f(-kh) = \int_0^\infty e^{-khu} \, d\sigma_h(u) \quad (k = 0, 1, 2, ...),$$

where $\sigma_h(u)$ is a certain non-decreasing function of u, the total variation of which is

$$\int_0^\infty d\sigma_h(u) = f(0) < \infty.$$

We now let h traverse the sequence $\{2^{-n}\}_1^\infty$. Again, as in many questions discussed previously we have arrived at a case when Helly's theorem has to be applied. Leaving out some simple arguments, we formulate the result: there exists a bounded non-decreasing function $\sigma(u)$ for which

$$f(-r) = \int_0^\infty e^{-ru} \, d\sigma(u), \qquad ...[5.35]$$

where r ranges over the aggregate of all non-negative rational fractions with denominators of the form 2^n ($n = 1, 2, ...$). Since for $x \leqq 0$ the two functions

$$f(x), \quad \int_0^\infty e^{xu} \, d\sigma(u)$$

are continuous, it follows from equation [5.35] that

$$f(x) = \int_0^\infty e^{xu} \, d\sigma(u) \quad (x \leqq 0).$$

5.2. We now amplify the results of the previous section in certain directions.

In the first place we note that the absolutely monotonic function $f(x)$ ($-\infty < x < 0$) represents the values on the semi-axis $-\infty < x < 0$ of an analytic function regular in the half plane $\operatorname{Re} z < 0$.

Indeed, for $f(0) < \infty$ this follows directly from the integral representation [5.33]. If however $f(0) = \infty$ one can introduce the function $f_\delta(x) = f(-\delta + x)$ where $\delta > 0$ is an arbitrarily small constant. In such a case we have, according to Theorem **5.5.1** which has already been proved,

$$f(x - \delta) = \int_0^\infty e^{xu} \, d\tau_\delta(u) \quad (-\infty < x \leqq 0),$$

where $\tau_\delta(u)$ is a non-decreasing function of bounded variation. Hence our statement follows for the half-plane $\operatorname{Re} z < -\delta$ and it remains to take into account that $\delta > 0$ is arbitrary.

From the fact just proved it follows that a function which is absolutely monotonic on the semi-axis $-\infty < x < 0$ can be defined as one for which all derivatives exist and

$$f^{(k)}(x) \geqq 0 \quad (k = 0, 1, 2, ...; \quad -\infty < x < 0).$$

This fact was proved by Bernstein without using an integral representation and for functions in any interval, not only on the semi-axis.

Further we note that in the case when $f(0) < \infty$ the function $f(z)$ is continuous right up to the imaginary axis ($x = 0$) and that the equality

$$f(x + iy) = \int_0^\infty e^{iyu} e^{xu} \, d\sigma(u)$$

holds. This equality shows that the values of the function $f(z)$ on any straight line $x = \operatorname{const} \leqq 0$, parallel to the imaginary axis represent a hermitian-positive function. Hence, incidentally, it follows that in equation [5.33] the function $\sigma(u)$ is determined by $f(x)$ essentially uniquely.

The connection with hermitian-positive functions is not limited to this. As was noted by SCHOENBERG[1], if $f(x)$ is absolutely monotonic for $-\infty < x < 0$ and if $f(0) < \infty$, then $f(-x^2) \in R(E_\infty)$.

From the remarks made it is easy to obtain:

THEOREM 5.5.2 (WIDDER[1]). *The function*

$$f(x) \quad (-\infty < x < 0)$$

can be represented in the form

$$f(x) = \int_0^\infty e^{xu} \, d\sigma(u),$$

where $\sigma(u)$ is a non-decreasing function, if and only if $f(x)$ is absolutely monotonic for $-\infty < x < 0$.

Proof. The distinction from Theorem 5.5.1 consists in the fact that here $x < 0$ and not $x \leq 0$ and, in connection with this, we do not require that the inequality $f(-0) < \infty$ should hold. We shall prove only what requires proving beyond 5.5.1.

It was remarked above that for any $\delta > 0$ one can write down the representation

$$f(x-\delta) = \int_0^\infty e^{xu} \, d\tau_\delta(u) \quad (-\infty < x \leq 0).$$

Hence

$$f(x) = \int_0^\infty e^{xu} \, d\sigma_\delta(u) \quad (-\infty < x \leq -\delta),$$

where

$$d\sigma_\delta(u) = e^{\delta u} \, d\tau_\delta(u),$$

so that $\sigma_\delta(u)$ is again a non-decreasing function, but in contrast to $\tau_\delta(u)$ it may be of unbounded variation. The function $\sigma_\delta(u)$, as was remarked above, can be determined from the function $f(x)$ essentially uniquely. Therefore $\sigma_\delta(u)$ does not depend on δ and so

$$f(x) = \int_0^\infty e^{xu} \, d\sigma(u),$$

where x can be any negative number.

5.3. DEFINITION 5.5.3 (SCHOENBERG[1]).   The function $g(x)$ $(-\infty < x < 0)$ belongs to the *class* T if $g(x) < 0$ $(-\infty < x < 0)$ and $g'(x) \in \mathfrak{M}(-\infty, 0)$.

THEOREM 5.5.3 (SCHOENBERG[1]). *The function*

$$g(x)   (-\infty < x < 0)$$

belongs to the class T *if and only if*

$$g(x) = c_0 + c_1 x + \int_{+0}^{\infty} \frac{e^{xu} - 1}{u} \, d\sigma(u)   (-\infty < x < 0), \quad ...[5.36]$$

where $c_0 \leqq 0$ and $c_1 \geqq 0$ are constants and $\sigma(u)$ is a non-decreasing function.

Proof. The only non-trivial part is the proof of the fact that from $g(x) \in$ T the representation [5.36] follows. This proof will now be given. Let $g(x) \in$ T. In that case

$$g'(x) = \int_0^{\infty} e^{xu} \, d\sigma(u)   (-\infty < x < 0).$$

Hence

$$g(x) - g(-\varepsilon) = \int_0^{\infty} \frac{e^{xu} - e^{-\varepsilon u}}{u} \, d\sigma(u)$$

$$= (x + \varepsilon)\{\sigma(+0) - \sigma(0)\} + \int_{+0}^{\infty} \frac{e^{xu} - e^{-\varepsilon u}}{u} \, d\sigma(u). \quad ...[5.37]$$

We fix $\varepsilon > 0$ and assume that $-\varepsilon < x < 0$. Then

$$\int_a^b \frac{e^{xu} - e^{-\varepsilon u}}{u} \, d\sigma(u) \leqq g(x) - g(-\varepsilon)   (b > a > 0),$$

whence

$$\int_a^b \frac{1 - e^{-\varepsilon u}}{u} \, d\sigma(u) \leqq g(-0) - g(-\varepsilon)$$

and so

$$\int_{+0}^{\infty} \frac{1 - e^{-\varepsilon u}}{u} \, d\sigma(u) \leqq g(-0) - g(-\varepsilon).$$

Therefore

$$\lim_{\varepsilon \to 0} \int_{+0}^{\infty} \frac{1 - e^{-\varepsilon u}}{u} \, d\sigma(u) = 0.$$

This relation shows that one can go to the limit $\varepsilon \to 0$ in equation [5.37], as a result of which one obtains the representation [5.36] with $c_1 = \sigma(+0) - \sigma(0)$.

The part played by the class T shows up in the answer to the following question: *what conditions must the function $g(x)$ ($-\infty < x < 0$) satisfy in order that the inclusion $f(x) \in \mathfrak{M}(-\infty, 0)$ should imply the inclusion $f[g(x)] \in \mathfrak{M}(-\infty, 0)$?*

We show that the required *necessary and sufficient condition reduces to the demand that $g(x) \in$ T.* The sufficiency may be verified by simple differentiation of the compound function

$$F(x) = f(y), \quad y = g(x):$$

$$F(x) = f(y) \geqq 0, \quad F'(x) = f'(y)g'(x) \geqq 0,$$

$$F''(x) = f'(y)g''(x) + f''(y)[g'(x)]^2 \geqq 0, \dots.$$

In order to prove the necessary condition we take a single function namely $f(x) = e^{vx}$ where $v > 0$ is a parameter.

The necessity of the condition $g(x) < 0$ is evident. Further, we must have the inequality

$$\frac{d}{dx} f[g(x)] \geqq 0$$

or

$$ve^{vy}g'(x) \geqq 0.$$

it follows from this that $g'(x) \geqq 0$. Differentiating once again, we obtain

$$ve^{vy}\{g''(x) + v[g'(x)]^2\} \geqq 0,$$

whence

$$g''(x) + v[g'(x)]^2 \geqq 0. \qquad \dots[5.38]$$

Further differentiation leads to inequalities of the form

$$ve^{vy}\{g^{(n)}(x) + vP_n[v, g'(x), g''(x), \dots, g^{(n-1)}(x)]\} \geqq 0,$$

where $P_n[\dots]$ is a polynomial in its arguments and n ranges over the values $3, 4, 5, \dots$ It follows from the last inequality that

$$g^{(n)}(x) + vP_n[v, g'(x), g''(x), \dots, g^{(n-1)}(x)] \geqq 0. \qquad \dots[5.39]$$

Now it remains to let the parameter v go to zero in [5.38] and [5.39], by which means we find that all derivatives $g^{(n)}(x)$ must be $\geqq 0$.

5.4. In conclusion we consider a special sub-class of convex functions in a given interval (a, b) which were introduced by BERNSTEIN[2]. We denote this sub-class by $W_{a, b}$.

DEFINITION 5.5.4.   A function $f(x)$ $(a < x < b)$ belongs to the *class* $W_{a,\,b}$ if it is continuous† and if all forms

$$\sum_{i,\,k=1}^{n} f(x_i+x_k)\zeta_i\zeta_k \quad (n = 1, 2, 3, \ldots)$$

are non-negative, assuming of course that all sums x_i+x_k are within the interval (a, b).

Writing down the fact that the form is non-negative for $n = 1$, we find that $f(x) \geqq 0$. Then, taking $n = 2$ we obtain the inequality

$$f(x_1)\zeta_1^2 + 2f\left(\frac{x_1+x_2}{2}\right)\zeta_1\zeta_2 + f(x_2)\zeta_2^2 \geqq 0. \qquad \ldots[5.40]$$

Hence, for $\zeta_1 = -1$, $\zeta_1 = 1$, it follows that‡

$$f\left(\frac{x_1+x_2}{2}\right) \leqq \frac{f(x_1)+f(x_2)}{2},$$

i.e. a function $f(x)$ of the class $W_{a,\,b}$ does indeed satisfy the condition of convexity.   It follows from inequality $[5.40]$ also that

$$f\left(\frac{x_1+x_2}{2}\right) \leqq \sqrt{f(x_1)f(x_2)},$$

whence one has the result that a function $f(x)$ of the class $W_{a,\,b}$ is, in addition, also logarithmically convex, i.e. its logarithm is a convex function.

Bernstein calls functions $f(x) \in W_{a,\,b}$ *exponentially convex*. This term is justified by an integral representation for any function $f(x) \in W_{a,\,b}$ to which we now turn.

We note that independently of Bernstein, but somewhat later, WIDDER[1] also introduced the class $W_{a,\,b}$ and studied it.

Both these authors proved the following:

† This requirement may be relaxed. See the following footnote.
‡ If the function $f(x)$ $(a < x < b)$ is measurable and finite almost everywhere, if in addition

$$-\infty < f(x) \leqq \infty \quad (a < x < b)$$

and if

$$f\left(\frac{x_1+x_2}{2}\right) \leqq \frac{f(x_1)+f(x_2)}{2} \quad (a < x_1, x_2 < b),$$

then $f(x)$ is continuous. This theorem is due to SIERPINSKI[1] (see also HIRSCHMAN and WIDDER[1], p. 122). It follows from the theorem that in the definition of the class $W_{a,\,b}$ we could have required the properties of measurability and finiteness almost everywhere in place of continuity.

THEOREM 5.5.4. *In order that the representation*

$$f(x) = \int_{-\infty}^{\infty} e^{xu} \, d\sigma(u) \quad (a < x < b)$$

be valid, where $\sigma(u)$ *is a non-decreasing function, it is necessary and sufficient that* $f(x) \in W_{a,\,b}$.

The proof of necessity presents no difficulty and can be omitted. We turn to the proof of sufficiency.

Thus let $f(x) \in W_{a,\,b}$. We take any closed interval $[\alpha, \beta] \subset (a, b)$ and choose a positive integer m. Then taking $n = 2^m - 1$, $h = 1/2^m$ and $l = \beta - \alpha$ we can state that

$$\sum_{i,\,k=0}^{n} f(\alpha + (i+k)h)\xi_i \xi_k \geqq 0,$$

$$\sum_{i,\,k=0}^{n-1} f(\alpha + h + (i+k)h)\xi_i \xi_k \geqq 0.$$

These relations are necessary and sufficient conditions for the solubility of a certain truncated Stieltjes moment problem and therefore there exists a non-decreasing function $\omega_n(t)$ such that

$$f(\alpha + kh) = \int_{0}^{\infty} t^k \, d\omega_n(t) \quad (k = 0, 1, 2, \ldots, 2^m - 1).$$

Putting $t = e^{hu}$ we rewrite these equations in the form

$$f(\alpha + kh) = \int_{-\infty}^{\infty} e^{khu} \, d\tau_n(u) \quad (k = 0, 1, 2, \ldots, 2^m - 1),$$

where

$$\tau_n(u) = \omega_n(e^{hu}).$$

The variation of each of the functions $\tau_n(u)$ is equal to $f(\alpha)$. Therefore we can apply Helly's theorem and, thus find a non-decreasing function $\tau(u)$ and a certain sub-sequence $\{n_i\}_1^{\infty}$ such that at all points of continuity of $\tau(u)$ we have

$$\lim_{i \to \infty} \tau_{n_i}(u) = \tau(u).$$

We show that at every point $r = p2^{-q}l$, where p and q are integers with $0 < p2^{-q} < 1$, the following equation is true:

$$f(\alpha + r) = \int_{-\infty}^{\infty} e^{ru} \, d\tau(u). \qquad \ldots [5.41]$$

To do this we take two points $r' = p'2^{-q}l$ and $r'' = p''2^{-q}l$ of the same form, such that $r' < r < r''$ and we hold these fixed. Then we assume that

n ranges over the sequence $\{n_i\}_1^\infty$. In this case, if we take as $-A < 0$, $B > 0$ points of continuity of the function $\tau(u)$ we get, for any sufficiently large n

$$f(\alpha + r) = \int_{-\infty}^{\infty} e^{ru} \, d\tau_n(u)$$

$$= \int_{-A}^{B} e^{ru} \, d\tau_n(u) + \int_{-\infty}^{-A} e^{ru} \, d\tau_n(u) + \int_{B}^{\infty} e^{ru} \, d\tau_n(u).$$

In the first integral on the right hand side we can go to the limit, on the other side we have

$$\int_{B}^{\infty} e^{ru} \, d\tau_n(u) = \int_{B}^{\infty} e^{r''u} e^{-(r''-r)u} \, d\tau_n(u)$$

$$\leqq e^{-(r''-r)B} \int_{B}^{\infty} e^{r''u} \, d\tau_n(u) \leqq e^{-(r''-r)B} f(\alpha + r'')$$

and analogously

$$\int_{-\infty}^{-A} e^{ru} \, d\tau_n(u) \leqq e^{-(r-r')A} f(\alpha + r').$$

Therefore

$$\left| f(\alpha + r) - \int_{-A}^{B} e^{ru} \, d\tau(u) \right| \leqq e^{-(r-r')A} f(\alpha + r') + e^{-(r''-r)B} f(\alpha + r'')$$

and, letting A and B go to infinity we get relation [5.41]. As both sides are continuous this equation is true for any r in the interval $[0, \beta - \alpha]$. Therefore

$$f(x) = \int_{-\infty}^{\infty} e^{xu} e^{-\alpha u} \, d\tau(u) = \int_{-\infty}^{\infty} e^{xu} \, d\sigma(u) \quad (\alpha \leqq x \leqq \beta), \qquad \ldots [5.42]$$

where $\sigma(u)$ is a non-decreasing function. The representation we have obtained shows that $f(x)$ is the value of an analytic function $f(z)$ regular in the strip $a < \operatorname{Re} z < b$. Hence, as was done once before, we can conclude that the representation [5.42] is true not only for $\alpha \leqq x \leqq \beta$, but also in the whole open interval $a < x < b$, and also that the function $\sigma(u)$ in this representation is essentially unique.

Addenda and Problems

1. Let $f(x)$ $(-a < x < a)$ be a continuous positive even function which in the interval $[0, a)$ is convex and decreasing. In this case $f(x)$

allows an infinitely large number of different continuations to functions of the class P_∞ (and therefore $f(x) \in P_a$).

Proof. The function $f(x)$ can be continued in an infinite variety of ways on to the whole of the number axis in such a way that the function $F(x)$ obtained is even, positive, convex for $x \geqq 0$ and tending monotonically to zero when $x \to \infty$. It is well known that then (see e.g., TITCHMARSH[1]) for real $t \neq 0$ the integral

$$p(t) = \frac{1}{\pi} \int_0^\infty F(x) \cos xt \, dx$$

has a meaning with

$$p(t) \geqq 0, \quad \int_0^\infty p(t) \, dt = \lim_{\varepsilon \to 0} \int_\varepsilon^\infty p(t) \, dt < \infty$$

and that the representation

$$F(x) = \int_{-\infty}^\infty p(t) e^{itx} \, dt$$

holds, whence it follows that $F(x) \in P_\infty$ and therefore $f(x) \in P_a$.

According to a verbal communication from Krein this simple fact was known to him in 1939 and served as the first impetus for investigating the continuation of hermitian-positive functions.

The first example of a non-unique continuation of a hermitian-positive function is due to GNEDENKO[1]. In this example he takes the function

$$f_1(x) = \begin{cases} 1 - 3 \, |x| + 2 \, |x|^3 & (-1 \leqq x \leqq 1), \\ 0 & (|x| > 1), \end{cases}$$

which permits the representation

$$f_1(x) = \int_{-\infty}^\infty e^{itx} \frac{6 \left(t \cos \dfrac{t}{2} - 2 \sin \dfrac{t}{2} \right)^2}{\pi t^4} \, dt$$

and therefore is hermitian-positive. On the other hand he constructs the periodic function $f_2(x)$ defined by the equations

$$f_2(x) = f_1(x) \quad (-1 \leqq x \leqq 1),$$

$$f_2(x+2) = f_2(x).$$

It permits the expansions

$$f_2(x) = \frac{3}{\pi^2} \sum_{n=1}^\infty \frac{\cos 2\pi nx}{n^2} + \frac{48}{\pi^4} \sum_{n=1}^\infty \frac{\cos (2n-1)\pi x}{(2n-1)^4}$$

and therefore is also hermitian-positive. In the interval $[-1, 1]$ the functions $f_1(x)$ and $f_2(x)$ coincide.

2.† By the class G_a $(a \leqq \infty)$ one means the aggregate of all functions $g(x)$ $(-a < x < a)$ which are continuous at the point $x = 0$, satisfy the conditions

$$g(0) = 0, \quad g(-x) = \overline{g(x)}$$

and generate an hermitian-positive kernel $K(x, y)$ $(0 \leqq x, y < a)$ according to the formula

$$K(x, y) = g(x) + g(-y) - g(x-y).$$

It is to be proved that any function $g(x) \in G_a$ possesses the following properties:

1) For $-\tfrac{1}{2}a < x < \tfrac{1}{2}a$ the inequality

$$| g(2x) | \leqq M | g(x) |$$

holds where M is an absolute constant $(M < 8)$,

2) For any $x_1, x_2 \in [0, a)$ one has the inequality

$$| g(x_2) - g(x_1) |$$

$$\leqq 2 \left| \operatorname{Im} g\left(\frac{x_2 - x_1}{2}\right) \right| + 2 \sqrt{\left| \operatorname{Re} g\left(\frac{x_1 + x_2}{2}\right) \right|} \cdot | \operatorname{Re} g(x_2 - x_1) |.$$

Hint. Write down the inequality

$$\sum_{j, k=1}^{n} K(x_j, x_k)\xi_j \bar{\xi}_k \geqq 0$$

first for $n = 2$, $x_1 = x$, $x_2 = 2x$ $(0 < x < \tfrac{1}{2}a)$ and then for $n = 3$, $x_3 = \tfrac{1}{2}(x_1 + x_2)$, $\xi_1 = \xi$, $\xi_2 = -\xi$, $\xi_3 = \eta$.

3. If $g(x) \in G_\infty$ and $M_\delta = \sup_{0 \leqq x \leqq \delta} | g(x) |$, then

1) For any $x \in (-\infty, \infty)$ the inequality holds:

$$| g(x) | \leqq M_\delta \left(1 + \frac{8 | x |^3}{\delta^3}\right),$$

2) For any $h \geqq 0$ and $x - h \geqq 0$ the inequality holds:

$$| g(x+h) - g(x-h) | \leqq 2 | \operatorname{Im} g(h) | + 2\sqrt{M_\delta \left(1 + \frac{8x^3}{\delta^3}\right)} \cdot | \operatorname{Re} g(2h) |.$$

† In composing examples 2–9 the article by AKHIEZER and GLAZMAN[2] has been used.

4. If $f(x) \in P_a$ then $f(0) - f(x) \in G_a$.

Proof. We must verify that the kernel

$$f(0) - f(x) - f(-y) + f(x-y)$$

is hermitian-positive. For this purpose we write down the identity

$$\sum_{\alpha,\,\beta=1}^{2n} f(z_\alpha - z_\beta)\zeta_\alpha\bar{\zeta}_\beta$$

$$= \sum_{p,\,q=1}^{n} f(z_{2p} - z_{2q-1})\zeta_{2p}\bar{\zeta}_{2q-1} + \sum_{p,\,q=1}^{n} f(z_{2p} - z_{2q})\zeta_{2p}\bar{\zeta}_{2q}$$

$$+ \sum_{p,\,q=1}^{n} f(z_{2p-1} - z_{2q-1})\zeta_{2p-1}\bar{\zeta}_{2q-1} + \sum_{p,\,q=1}^{n} f(z_{2p-1} - z_{2q})\zeta_{2p-1}\bar{\zeta}_{2q}.$$

In this identity we put $z_{2p} = 0$, $z_{2p-1} = x_p$, $\zeta_{2p-1} = \xi_p$, $\zeta_{2p} = -\xi_p$ ($p = 1, 2, ..., n$) and get

$$0 \leqq \sum_{\alpha,\,\beta=1}^{2n} f(z_\alpha - z_\beta)\zeta_\alpha\bar{\zeta}_\beta = \sum_{p,\,q=1}^{n} \{f(0) - f(x_p) - f(-x_q) + f(x_p - x_q)\}\xi_p\bar{\xi}_q.$$

5. If $g(x) \in G_a$, then $f(x) \equiv \exp\{-g(x)\} \in P_a$ (KREIN[3]).

Hint: Use SCHUR'S[1] theorem† which states that the positive nature of the forms

$$\sum_{p,\,q=1}^{n} a_{pq}\xi_p\bar{\xi}_q, \qquad \sum_{p,\,q=1}^{n} b_{pq}\xi_p\bar{\xi}_q$$

implies the positive property for the form

$$\sum_{p,\,q=1}^{n} a_{pq}b_{pq}\xi_p\bar{\xi}_q.$$

6. A function $F(x)$ ($-a < x < a$) belongs to the class P_a^* ($a \leqq \infty$) if for every positive integer n it is the n-th power of some function of the class P_a (thus $P_a^* \subset P_a$).

The class P_∞^* with the normalization $F(0) = 1$ plays an important part in the theory of probability. It represents the aggregate of characteristic functions of so-called *infinitely divisible distribution laws*.

If $g(x) \in G_a$ then $F(x) \equiv \exp\{-g(x)\} \in P_\infty^*$ (see KREIN[3]).

7. For any function $F(x) \in P_a^*$ one can find such a function $g(x) \in G_\infty$, that

$$F(x) = F(0)\exp\{-g(x)\} \quad (-a < x < a).$$

† See also POLYA and SZEGÖ[1].

Proof. Without destroying generality we assume that $F(0) = 1$. By virtue of the definition of the class P_a^* one can find a function $f_n(x) \in P_a$ for any positive integer n such that

$$F(x) = [f_n(x)]^n \quad (-a < x < a).$$

The function $f_n(x)$ can be continued into a certain function $\tilde{f}_n(x) \in P_\infty$. With this $\tilde{f}_n(x)$ we construct the function

$$g_n(x) = n\{1 - \tilde{f}_n(x)\},$$

which, according to example 4, belongs to the class G_∞.

We choose a positive number $\delta < \frac{1}{2}a$ which is so small that for $-\delta \leqq x \leqq \delta$ the value of $F(x)$ is sufficiently close to 1. Putting

$$\Phi(x) = \operatorname{Im} \ln F(x),$$

we have for $-\delta \leqq x \leqq \delta$ the following relation:

$$\operatorname{Re}\left[g_n(x)\right] = n \operatorname{Re}\left\{1 - [F(x)]^{\frac{1}{n}}\right\} = n\left\{1 - |F(x)|^{\frac{1}{n}} \cos \frac{\Phi(x)}{n}\right\},$$

$$\operatorname{Im}\left[g_n(x)\right] = -n |F(x)|^{\frac{1}{n}} \sin \frac{\Phi(x)}{n}.$$

Hence

$$\left| \operatorname{Re}\left[g_n(x)\right] \right| \leqq \ln \frac{1}{|F(x)|} + \frac{1}{2n}[\Phi(x)]^2,$$

$$\left| \operatorname{Im}\left[g_n(x)\right] \right| \leqq |\Phi(x)|.$$

With the aid of these estimates we establish that

$$M_\delta^{(n)} \equiv \max_{0 \leqq x \leqq \delta} |g_n(x)| \leqq \omega(\delta),$$

and also that for $-h \leqq \varepsilon \leqq h \ (h \leqq \delta)$

$$\left| \operatorname{Re}\left[g_n(\varepsilon)\right] \right| + \left| \operatorname{Im}\left[g_n(\varepsilon)\right] \right| \leqq \omega(h),$$

where the quantity

$$\omega(h) = \sup_{|x| \leqq h} \left\{\ln \frac{1}{|F(x)|} + |\Phi(x)| + \frac{1}{2}[\Phi(x)]^2\right\}$$

is independent of n and tends to zero with h.

Therefore, using example 3 the sequence of functions $g_n(x)$ $(-\infty < x < \infty, \ n = 1, 2, 3, \ldots)$ is uniformly bounded and equicontinuous in every finite interval. Therefore there exists such a sequence $\{g_{n_k}(x)\}_{k=1}^{\infty}$ and such a continuous function $g(x)$ $(-\infty < x < \infty)$ that the limit

$$\lim_{k \to \infty} g_{n_k}(x) = g(x)$$

is uniform in every finite interval. This limiting function $g(x)$ evidently belongs to the class G_{∞}; on the other hand we have in the interval $(-a, a)$:

$$g(x) = \lim_{k=\infty} n_k\{1 - [F(x)]^{\frac{1}{n_k}}\} = \ln F(x).$$

8. Any function $F(x) \in P_a^*$ $(a < \infty)$ can be continued into a function of the class P_{∞}^*.

Any function $g(x) \in G_a$ $(a < \infty)$ can be continued into a function of the class G_{∞} (KREIN[3]).

9. Any function $g(x) \in G_a$ $(a \leq \infty)$ permits the representation

$$g(x) = i\beta x + \int_{-\infty}^{\infty} \left\{1 + \frac{iux}{1+u^2} - e^{iux}\right\} \frac{d\omega(u)}{u^2} \quad (-a < x < a), \quad \dots [5.43]$$

where β is a real constant and $\omega(u)$ is a non-decreasing function for which

$$\int_{-\infty}^{\infty} \frac{d\omega(u)}{1+u^2} < \infty.$$

Here the constant is determined uniquely by the function $g(x)$ and for $a = \infty$ the function $\omega(u)$ is determined essentially uniquely.

Conversely, any function $g(x)$ which permits the representation [5.43] belongs to the class G_a (KREIN[3]).

Proof. The second statement can be verified without any difficulty. As for the first statement, it follows from example **8** that it is sufficient to prove it for the case $a = \infty$. In that case example **6** ensures that the question reduces to obtaining an integral representation of a function of the class P_{∞} which has long been known from the investigations of P. LÉVY[1] and KHINCHIN[1].

It is in any case simple to give a direct proof of equation [5.43] for a function $g(x) \in G_{\infty}$. To do this we note that

$$|g(x)| \leq A + B|x|^3.$$

Therefore the Laplace transform

$$H(\zeta) = \zeta^2 \int_0^{\infty} g(x)e^{ix\zeta} dx \quad \dots [5.44]$$

has a meaning in the half plane $\eta = \operatorname{Im} \zeta > 0$ and gives an analytic function. With the aid of simple transformations one establishes that

$$\frac{H(\zeta)+\overline{H(\zeta)}}{2\eta} = \zeta\bar{\zeta}\int_0^\infty \int_0^\infty \{g(\alpha-\beta)-g(\alpha)-g(-\beta)\}e^{i\alpha\zeta}e^{-i\beta\bar\zeta}\,d\alpha\,d\beta.$$

Since $g(x) \in G_\infty$, the right hand side is negative, therefore $i^{-1}H(\zeta) \in N$. and therefore $H(\zeta)$ can be represented in the form

$$H(\zeta) = i\alpha\zeta - i\beta - i\int_{-\infty}^\infty \frac{1-u\zeta}{u+\zeta}\frac{d\omega(u)}{1+u^2},$$

where $\omega(u)$ is a non-decreasing function for which

$$\int_{-\infty}^\infty \frac{d\omega(u)}{1+u^2} < \infty,$$

and $\alpha \geq 0$, and $\beta \geq 0$ are two constants.

Noting that

$$\frac{1}{i}\frac{u^2(1-u\zeta)}{(u+\zeta)(1+u^2)} = \zeta^2\int_0^\infty \left\{1+\frac{iux}{1+u^2}-e^{iux}\right\}e^{ix\zeta}\,dx,$$

we obtain the representation

$$H(\zeta) = i\alpha\zeta - i\beta + \zeta^2\int_0^\infty e^{ix\zeta}\,dx\int_{-\infty}^\infty \left\{1+\frac{iux}{1+u^2}-e^{iux}\right\}\frac{d\omega(u)}{u^2}.$$

It remains to compare this representation with [5.44] and to note that

$$1 = -i\zeta\int_0^\infty e^{ix\zeta}\,dx = -\zeta^2\int_0^\infty xe^{ix\zeta}\,dx \quad (\operatorname{Im} \zeta > 0).$$

10. The function $S(x)$ $(0 \leq x < a;\ a \leq \infty)$ belongs to the class K_a if it is continuous, becomes zero for $x = 0$ and generates an hermitian-positive kernel according to the formula

$$K(x, y) = \tfrac{1}{2}\{S(x+y)-S(|x-y|)\} \quad \left(0 \leq x, y < \frac{a}{2}\right). \quad \dots[5.45]$$

Any function $S(x) \in K_a$ allows the representation

$$S(x) = \int_{-\infty}^\infty \frac{1-\cos(x\sqrt{t})}{t}\,d\sigma(t), \quad \dots[5.46]$$

where $\sigma(t)$ is a non-decreasing function.

Conversely any function $S(x)$ $(0 \leq x < a)$ which allows such a representation belongs to the class K_a.

This theorem was established by KREIN[6] and was initially just an example of the application of the method of directing functionals. Later on the rôle of the class K_∞ became clear for certain new problems† relating to the Sturm-Liouville equation on the semi-axis.

The second statement of the theorem may be proved by examining the integral

$$\int_{-\infty}^{\infty}\left\{\sum_{k=1}^{m}\xi_k\frac{\sin(x_k\sqrt{t})}{\sqrt{t}}\right\}^2 d\sigma(t)\geqq 0.$$

The first statement can be obtained most simply with the aid of the method of directing functionals.

With this aim we take as our starting point the real linear manifold G of all functions

$$\phi(s)=\int_{\alpha}^{\beta}\frac{\sin(x\sqrt{s})}{\sqrt{s}}d\mu(x),$$

where $[\alpha,\beta]$ is an arbitrary interval contained in the interval $[0,\tfrac{1}{2}a)$ and $\mu(x)$ is any real function of bounded variation. Further, starting from the equation

$$\left(\frac{\sin(x\sqrt{s})}{\sqrt{s}},\frac{\sin(y\sqrt{s})}{\sqrt{s}}\right)=K(x,y)\quad\left(0\leqq x,y<\frac{a}{2}\right),$$

we perform a metrization of G and then, by a well-known procedure, construct a Hilbert space H on the base G. Of particular importance is the manifold D of functions $\psi(s)\in G$ which may be represented in the form

$$\psi(s)=\int_0^c\frac{\sin(x\sqrt{s})}{\sqrt{s}}\gamma(x)\,dx\quad\left(0<c<\frac{a}{2}\right),$$

where $\gamma(x)$ has an absolutely continuous first derivative and satisfies the conditions $\gamma(0)=\gamma(c)=\gamma'(c)=0$. We prove that the set D is dense in G. To do this it is sufficient to verify that the function $\phi_0(s)=[\sin(c\sqrt{s})]/\sqrt{s}$, where $0<c<\tfrac{1}{2}a$ can be sufficiently well-approximated by a function from D. We put

$$\psi_0(s)=\int_0^{c+h}\frac{\sin(x\sqrt{s})}{\sqrt{s}}\gamma_0(x)\,dx,$$

† For these problems see MARCHENKO[1], GEL'FAND and LEVITAN[1] and KREIN[13]. Functions of the form [5.46] have an interesting and useful mechanical interpretation which was discovered by Krein.

where $\gamma_0(x) = 0$ for $x \leqq c - h$ and

$$\int_{c-h}^{c+h} \gamma_0(x)\, dx = 1.$$

For instance, we can put

$$\gamma_0(x) = \frac{15}{16} \frac{[h^2 - (x-c)^2]^2}{h^5} \quad (c - h \leqq x \leqq c + h).$$

In such a case we have

$$\theta_0(s) \equiv \phi_0(s) - \psi_0(s) = \int_{c-h}^{c+h} \left\{ \frac{\sin (c\sqrt{s})}{\sqrt{s}} - \frac{\sin (x\sqrt{s})}{\sqrt{s}} \right\} \gamma_0(x)\, dx,$$

and therefore

$(\theta_0(s), \theta_0(s))$

$$= \int_{c-h}^{c+h} \int_{c-h}^{c+h} \{K(c, c) - K(c, y) - K(x, c) + K(x, y)\} \gamma_0(x) \gamma_0(y)\, dx\, dy.$$

Because of the continuity of the kernel $K(x, y)$ this expression will be arbitrarily small for sufficiently small $h > 0$.

For A we take the operator of multiplication by the independent variable. If

$$\psi(s) = \int_0^c \frac{\sin (x\sqrt{s})}{\sqrt{s}} \gamma(x)\, dx \in D,$$

then by integration by parts we find that

$$s\psi(s) = -\int_0^c \frac{\sin (x\sqrt{s})}{\sqrt{s}} d\gamma'(x) \in G.$$

Therefore the domain of definition of the operator A contains D and therefore the operator A is symmetric.

It remains to show that the operator A has a directing functional. We prove that this functional is

$$\Phi(\phi, t) = \phi(t). \qquad \qquad \dots [5.47]$$

We can confine ourselves to discussing it in D, but then the verification of conditions a), b) and c) occurring in the definition of a directing functional (see Chapter 4 Section 3.4) does not present any difficulties.

Indeed the continuous differentiability of the function

$$\psi(t) = \int_0^c \frac{\sin (x\sqrt{t})}{\sqrt{t}} \gamma(x)\, dx \quad (-\infty < t < \infty)$$

is evident. To verify condition b) it is sufficient to take the function

$$v(s) = \int_0^h \frac{\sin (x\sqrt{s})}{\sqrt{s}} x^2 (h-x)^2 \, dx \in D,$$

for which the functional [5.47] will be different from zero in the interval $(-\infty, \pi^2/h^2]$, the right hand end of which can be taken arbitrarily large since the number $h > 0$ can be arbitrarily small.

To verify condition c) we consider the equation

$$A\chi - t\chi = \psi \qquad \qquad ...[5.48]$$

or

$$s\chi(s) - t\chi(s) = \psi(s),$$

where the function

$$\psi(s) = \int_0^c \frac{\sin (x\sqrt{s})}{\sqrt{s}} \gamma(x) \, dx \in D \quad \left(0 < c < \frac{a}{2}\right)$$

is given and the function $\chi(s)$ is to be found within the class D, namely in the form

$$\chi(s) = \int_0^c \frac{\sin (x\sqrt{s})}{\sqrt{s}} \theta(x) \, dx.$$

Using integration by parts, we find that

$$s\chi(s) - t\chi(s) = -\int_0^c \frac{\sin (x\sqrt{s})}{\sqrt{s}} \{\theta''(x) + t\theta(x)\} \, dx.$$

Thus it is sufficient to put

$$\theta''(x) + t\theta(x) = -\gamma(x).$$

Hence we find that

$$\theta(x) = -\int_0^x \frac{\sin \sqrt{t}(x-\xi)}{\sqrt{t}} \gamma(\xi) \, d\xi + B \frac{\sin (x\sqrt{t})}{\sqrt{t}} + B' \cos (x\sqrt{t})$$

and $B' = 0$ since $\theta(x)$ must become zero for $x = 0$. It is also required that $\theta(c) = \theta'(c) = 0$, These conditions can be written in the form

$$B \sin (c\sqrt{t}) - \int_0^c \gamma(\xi) \sin \sqrt{t}(c-\xi) \, d\xi = 0,$$

$$B \cos (c\sqrt{t}) - \int_0^c \gamma(\xi) \cos \sqrt{t}(c-\xi) \, d\xi = 0,$$

or

$$\sin (c\sqrt{t})\left\{B - \int_0^c \cos (\xi\sqrt{t})\gamma (\xi)\, d\xi\right\} + \cos (c\sqrt{t})\int_0^c \sin (\xi\sqrt{t})\gamma (\xi)\, d\xi = 0$$

and

$$\cos (c\sqrt{t})\left\{B - \int_0^c \cos (\xi\sqrt{t})\gamma (\xi)\, d\xi\right\} - \sin (c\sqrt{t})\int_0^c \sin (\xi\sqrt{t})\gamma (\xi)\, d\xi = 0.$$

They reduce to the following equations:

$$B = \int_0^c \cos (\xi\sqrt{t})\gamma (\xi)\, d\xi,$$

$$\int_0^c \frac{\sin (\xi\sqrt{t})}{\sqrt{t}}\, \gamma (\xi)\, d\xi = 0.$$

The first equation serves to determine the constant B, while the second represents a necessary and sufficient condition for equation [5.48] to be soluble within the class D. This condition just expresses the fact that the functional Φ must become zero for the function $\psi(s)$ at the point t.

Thus all conditions have been verified and Theorem 4.3.4 can be applied, i.e. one can write down the equation

$$(\psi_1(s), \psi_2(s)) = \int_{-\infty}^{\infty} \psi_1(t)\psi_2(t)\, d\sigma(t),$$

where $\psi_1(s)$ and $\psi_2(s)$ are arbitrary functions from the class D and $\sigma(t)$ is a certain non-decreasing function. Owing to the fact that the set D is dense in G this equation is true for any functions in G. Applying this result to the function

$$\frac{\sin (x\sqrt{s})}{\sqrt{s}}, \quad \frac{\sin (y\sqrt{s})}{\sqrt{s}} \quad \left(0 < x, y < \frac{a}{2}\right),$$

we find the representation

$$K(x, y) = \int_{-\infty}^{\infty} \frac{\sin (x\sqrt{t}) \cdot \sin (y\sqrt{t})}{t}\, d\sigma(t)$$

or

$$S(x+y) - S(|x-y|) = \int_{-\infty}^{\infty} \frac{\cos \sqrt{t}(x-y) - \cos \sqrt{t}(x+y)}{t}\, d\sigma(t),$$

whence for $y = x$ we get the required representation [5.46].

11. Prove the theorem of the previous example by M. Riesz' method.

Hint. We confine ourselves to explaining the scheme of the proof for $a < \infty$†.

We introduce the finite systems of numbers

$$R_n = \left\{\frac{a}{2^n}, \frac{2a}{2^n}, \frac{3a}{2^n}, ..., \frac{2^n - 1}{2^n} a\right\} \quad (n = 1, 2, 3, ...)$$

and denote by R their union. We denote by T(R) and correspondingly by $T(R_n)$ the real linear hull of all functions of u $(-\infty < u < \infty)$ of the form

$$\frac{1 - \cos(r\sqrt{u})}{u},$$

where r ranges over R, or correspondingly over R_n. Further, we introduce in T(R), and correspondingly in $T(R_n)$ the functional $\wp\{\phi\}$, which is linear in the algebraic sense, by putting

$$\wp\left\{\frac{1 - \cos(r\sqrt{u})}{u}\right\} = S(r).$$

It is easy to verify that this functional is non-negative. Further, if one considers the functional $\wp\{\phi\}$ in $T(R_n)$ it is non-negative in a wider sense, namely, in that for $\phi(u) \in T(R_n)$ the inequality $\wp\{\phi\} \geqq 0$ holds if only $\phi(u)$ satisfies the condition

$$\phi(u) \geqq 0 \quad \left(-\infty < u \leqq \frac{2^{2n}\pi^2}{a^2} = B_n\right).$$

We can apply Theorem **2.6.6** to the functional $\wp\{\phi\}$ in $T(R_n)$, taking for $\psi(u)$ the function

$$\psi(u) = \frac{1 - \cos\dfrac{a\sqrt{u}}{2^n}}{u} + \frac{1 - \cos\dfrac{(2^n - 1)a\sqrt{u}}{2^n}}{u}$$

the positive character of which within the interval $(-\infty, B_n)$ is obvious.

---

† If the theorem is proved for $a < \infty$, its proof for $a = \infty$ can be obtained with the aid of a limiting process based on Helly's theorems. Because of the infinite range of integration certain additional arguments are necessary which are similar to those used repeatedly in the preceding work.

On the basis of Theorem **2.6.6** there exists a non-decreasing function of bounded variation, $\sigma_n(u)$ $(-\infty \leqq u \leqq B_n)$ for which

$$S\left(\frac{ka}{2^n}\right) = \int_{-\infty}^{B_n} \frac{1 - \cos \frac{ka\sqrt{u}}{2^n}}{u} \, d\sigma_n(u) \quad (k = 0, 1, 2, \ldots, 2^n - 2).$$

Now we introduce the real functions

$$H_n(x) = \int_{-\infty}^{0} \frac{1 - \cos(x\sqrt{u})}{u} \, d\sigma_n(u),$$

$$(0 \leqq x < a),$$

$$G_n(x) = \int_{0}^{\infty} \frac{1 - \cos(x\sqrt{u})}{u} \, d\sigma_n(u)$$

where

$$\sigma_n(u) = \sigma_n(B_n) \quad (u \geqq B_n).$$

Each of the functions $H_n(x)$ and $G_n(x)$ becomes zero for $x = 0$ and is non-negative for all other x.

We apply Helly's first theorem to the sequence of functions $\{\sigma_n(u)\}_1^\infty$ $(-\infty \leqq u \leqq 0)$. This is possible because

$$\int_{-\infty}^{0} d\sigma_n(u) \leqq \frac{8}{a^2} S\left(\frac{a}{2}\right).$$

Then we apply Helly's second theorem, as the result of which we obtain the function

$$H(x) = \lim_{i \to \infty} H_{n_i}(x) = \int_{-\infty}^{0} \frac{1 - \cos(x\sqrt{u})}{u} \, d\sigma(u).$$

This function is continuous, it is zero at the point $x = 0$ and, whatever point $r \in R$ one takes, the equation

$$H(r) = H_n(r).$$

is true for any sufficiently large n. The function

$$G(x) = S(x) - H(x)$$

is also continuous and for any $r \in R$ and any sufficiently large n we have

$$G(r) = G_n(r).$$

Further, putting

$$G(-x) = G(x) \quad (0 \leqq x < a)$$

and using the integral representation of the function $G(x)$ we find that for any $r_k \in R$ $(k = 1, 2, ..., m)$ the form

$$\sum_{\alpha, \beta=1}^{m} \{G(r_\alpha) + G(r_\beta) - G(r_\alpha - r_\beta)\} \xi_\alpha \xi_\beta$$

is non-negative. By virtue of the fact that the function $G(x)$ is continuous, this property is conserved if one replaces the numbers r_α by any numbers $x_\alpha \in [0, a)$. Therefore $G(x)$ is a real function of the class G_a and thus permits the integral representation

$$G(x) = \int_0^\infty \frac{1 - \cos(x\sqrt{u})}{u} \, d\sigma(u) \quad (0 \leqq x < a),$$

where $\sigma(u)$ is a non-decreasing function, for which

$$\int_0^\infty \frac{d\sigma(u)}{1 + u} < \infty.$$

12. The problem of determining non-decreasing functions $\sigma(u)$ $(-\infty < u < \infty)$ from the function $S(x)$ $(0 \leqq x < a)$ with the aid of the equation

$$S(x) = \int_{-\infty}^\infty \frac{1 - \cos(x\sqrt{u})}{u} \, d\sigma(u)$$

can be considered as the continuum analogue of Hamburger's moment problem.

Purely formally, this statement can be explained as follows. We take the function

$$f(z) = \int_{-\infty}^\infty \frac{d\sigma(u)}{z - u},$$

where $\sigma(u)$ is a non-decreasing function. The integrand can be expressed as an expansion into a power series

$$\frac{1}{z - u} \sim \sum_{k=0}^\infty \frac{u^k}{z^{k+1}},$$

and also as a Laplace integral

$$\frac{1}{z - u} \sim i\sqrt{z} \int_0^\infty e^{ix\sqrt{z}} \frac{1 - \cos(x\sqrt{u})}{u} \, dx,$$

which is the continuous analogue of the power series. For $f(z)$ we shall obtain the following formal expansions

$$f(z) \sim \sum_{k=0}^{\infty} \frac{1}{z^{k+1}} \int_{-\infty}^{\infty} u^k \, d\sigma(u) = \sum_{k=0}^{\infty} \frac{s_k}{z^{k+1}},$$

$$f(z) \sim \int_{0}^{\infty} i\sqrt{z} e^{ix\sqrt{z}} \, dx \int_{-\infty}^{\infty} \frac{1 - \cos(x\sqrt{u})}{u} \, d\sigma(u)$$

$$= \int_{0}^{\infty} S(x) i\sqrt{z} e^{ix\sqrt{z}} \, dx.$$

In the first of these the coefficients are the moments

$$s_k = \int_{-\infty}^{\infty} u^k \, d\sigma(u),$$

and in the second their continuous analogue

$$S(x) = \int_{-\infty}^{\infty} \frac{1 - \cos(x\sqrt{u})}{u} \, d\sigma(u).$$

13. In m-dimensional Euclidean space E_m a measure $\omega(A)$ $(A \subset E_m)$ is given. Let $A_{x,u}$ $(x \neq 0)$ denote the set of all points $t = \{t_1, t_2, \ldots, t_m\} \in E_m$, for which

$$(t, x) \equiv t_1 x_1 + t_2 x_2 + \ldots + t_m x_m \leq u,$$

and let $\omega(A_{x,u}) = \sigma(u; x)$. It is to be proved that if for any vector $x \neq 0$ one knows the functions $\sigma(u; x)$ $(-\infty < u < \infty)$ one can determine the measure $\omega(A)$ (CRAMÉR and WOLD[1]).

Proof. In the space E_m we consider the hermitian-positive function

$$f(x) = \int_{E_m} e^{i(x, t)} \, d\omega(t), \qquad \ldots [5.49]$$

where $\omega(t)$ is a function of the point t, which is generated by the measure $\omega(A)$. Introducing in place of $t_1, t_2, \ldots, t_m$ new co-ordinates such that one of them is (t, x), we rewrite $[5.49]$ in the form

$$f(x) = \int_{-\infty}^{\infty} e^{iu} \, d_u \int_{(t, x) \leq u} d\omega(t) = \int_{-\infty}^{\infty} e^{iu} \, d_u \sigma(u; x).$$

Hence we see that if we know the family of functions $\sigma(u; x)$ $(-\infty < u < \infty)$ this allows us to determine the function $f(x)$ for $x \neq 0$. By continuity the function $f(x)$ may be determined also for $x = 0$. It then remains to use Theorem 5.4.1.

14. Let $\omega(A)$ $(A \subset E_m)$ be a measure in Euclidean space of $m > 1$ dimensions and $\omega(t)$ a point function of $t = \{t_1, t_2, ..., t_m\}$. generated by this measure. Assume that all the moments

$$s_{p_1 p_2 \, ... \, p_m} = \int_{E_m} t_1^{p_1} t_2^{p_2} \, ... \, t_m^{p_m} \, d\omega(t) \quad (p_i = 0, 1, 2, ...).$$

have finite values.

Prove that the moment problem expressed by these equations will certainly be determinate if

$$\sum_1^\infty \frac{1}{\sqrt[2k]{S_{2k}}} = \infty,$$

where

$$S_n = s_{n00 \, ... \, 0} + s_{0n0 \, ...} + \, ... \, + s_{000 \, ... \, n}$$

(CRAMÉR and WOLD[1]).

Proof. We introduce the function $\sigma(u; x)$ in the same way as in example **13** and put

$$s_k = s_k(x) = \int_{-\infty}^{\infty} u^k \, d_u \sigma(u; x) \quad (k = 0, 1, 2, ...). \qquad ...[5.50]$$

These "one-dimensional" moments may evidently be represented in the form

$$s_k(x) = \int_{E_m} (t_1 x_1 + t_2 x_2 + \, ... \, + t_m x_m)^k \, d\omega(t).$$

Since by the Cauchy-Hölder inequality we have

$$(t_1 x_1 + t_2 x_2 + \, ... \, + t_m x_m)^{2k} \leq |x|^{2k} (t_1^2 + t_2^2 + \, ... \, + t_m^2)^k$$

$$\leq |x|^{2k} (t_1^{2k} + t_2^{2k} + \, ... \, + t_m^{2k}) m^{k-1},$$

the estimates

$$s_{2k}(x) \leq m^{k-1} |x|^{2k} S_{2k},$$

are true and therefore, from the conditions of the theorem, we have

$$\sum_1^\infty \frac{1}{\sqrt[2k]{s_{2k}(x)}} = \infty.$$

Therefore we can use Carleman's criterion, (see Addenda and Problems to Chapter 2 example **11**) according to which the moment problem

$[5.50]$ is determinate for any $x \neq 0$. It then remains to use the results of example 13.

15. The two-dimensional† Hausdorff moment problem consists in the following: given numbers $s_{p,q}$ $(p, q = 0, 1, 2, ...)$ it is required to find a non-decreasing point function $\omega(u, v)$ $(0 \leq u \leq 1, \ 0 \leq v \leq 1)$ such that

$$s_{p,q} = \int_0^1 \int_0^1 u^p v^q \, d\omega(u, v) \quad (p, q = 0, 1, 2, ...).$$

Prove that:

a) This problem is soluble if and only if for any $m, n = 0, 1, 2, ...$ and any $p, q = 0, 1, 2, ...$ the inequality

$$\sum_{i=0}^m \sum_{k=0}^n (-1)^{i+k} \binom{m}{i} \binom{n}{k} s_{p+i, q+k} \geq 0$$

holds.

b) The solution is always unique (HILDEBRANDT and SCHOENBERG[1]).

16. **Absolutely monotonic functions of several variables.** Let the function $f(x) = f(x_1, x_2, ... x_m)$ be given, let it be continuous in the octant $x_1 \leq 0, \ x_2 \leq 0, ..., x_m \leq 0$ and let the inequality

$$\frac{\partial^{n_1 + n_2 + ... + n_m} f}{\partial x_1^{n_1} \partial x_2^{n_2} ... \partial x_m^{n_m}} \geq 0$$

hold at any internal point of this octant, for all combinations of numbers $n_1 \geq 0, n_2 \geq 0, ..., n_m \geq 0$.

Prove that under these conditions and only then, the representation

$$f(x) = \int_{D_m} e^{(x, t)} \, d\omega(t)$$

is valid where D_m is the first octant $(t_j \geq 0, j = 1, 2, ..., m)$ and $\omega(t)$ is a non-decreasing bounded point function in this octant. Prove also that the stated representation is unique.

(For this and certain more general propositions see the book by BOCHNER[3] where there is a reference to the dissertation by Gilbert, 1952).

---

† The case of an arbitrary number of dimensions differs only by some elaboration of notation.

17. Exponentially convex functions of many variables. Let a continuous function $f(x) = f(x_1, x_2, ..., x_m)$ be given in the interval $I_{a,b}$ of m-dimensional space and let all forms

$$\sum_{\alpha, \beta=1}^{N} f(x^\alpha + x^\beta)\xi_\alpha\xi_\beta \qquad ...[5.51]$$

be non-negative (it is of course assumed that all sums $x^\alpha + x^\beta$ belong to $I_{a,b}$). Prove that under these and only these conditions the representation

$$f(x) = \int_{E_m} e^{(x,t)} d\omega(t)$$

is valid, where $\omega(t)$ is a certain non-decreasing (possibly unbounded) point function in E_m. Prove also that this representation is unique.

The necessity of the condition and the uniqueness of the representation can be verified very simply.

We dwell on the proof of sufficiency. It can be obtained by induction. For $m = 1$ sufficiency has already been proved (Theorem 5.5.4). Therefore we must justify the passage from m to $m+1$. We shall do this only for $m = 1$ since for arbitrary m only the notation is more complicated and in essence the proof remains the same as for $m = 1$.

Thus we shall discuss the function $f(x) = f(x_1, x_2)$ and without reducing generality we assume that the interval I is symmetric with respect to zero, i.e. $x_1 \in (-\frac{1}{2}l_1, \frac{1}{2}l_1)$, $x_2 \in (-\frac{1}{2}l_2, \frac{1}{2}l_2)$. Owing to the fact that the point 0 now belongs to I the function $\omega(t)$ proves to be bounded.

We choose arbitrarily points $x_1^k \in (-\frac{1}{2}l_1, \frac{1}{2}l_1)$ $(k = 1, 2, ..., K)$ and points $x_2^p \in (-\frac{1}{2}l_2, \frac{1}{2}l_2)$ $(p = 1, 2, ..., P)$. Then we form KP vectors $\{x_1^k, x_2^p\}$, which will play the part of the vectors x^α in the form $[5.51]$.

For the corresponding coefficients we shall take the products $\xi_k\eta_p$. We shall then be able to write down the inequality

$$\sum_{k, k'=1}^{K} \sum_{p, p'=1}^{P} f(x_1^k + x_1^{k'}, x_2^p + x_2^{p'})\xi_k\xi_{k'}\eta_p\eta_{p'} \geqq 0.$$

If we put

$$g(x_1) = \sum_{p, p'=1}^{P} f(x_1, x_2^p + x_2^{p'})\eta_p\eta_{p'},$$

this inequality takes on the form

$$\sum_{k, k'=1}^{K} g(x_1^k + x_1^{k'})\xi_k\xi_{k'} \geqq 0. \qquad ...[5.52]$$

Since $g(x_1)$ is continuous, we thus find that it is exponentially convex.

In particular taking $P = 1$, we find that $f(x_1, x_2)$ is an exponentially convex function of x_1 for any fixed x_2†. Therefore one finds uniquely by Theorem **5.5.4** the representation

$$f(x_1, x_2) = \int_{-\infty}^{\infty} e^{x_1 t_1} \, d_{t_1} \sigma(t_1, x_2), \qquad \ldots [5.53]$$

where $\sigma(t_1, x_2)$ is a bounded non-decreasing function of t_1 which depends on the parameter x_2. We can here assume that

$$\sigma(-\infty, x_2) = 0, \quad \sigma(t_1 - 0, x_2) = \sigma(t_1, x_2).$$

With such a normalization $\sigma(t_1, x_2)$ is a continuous function of $x_2 \in (-\tfrac{1}{2}l_2, \tfrac{1}{2}l_2)$ for any fixed t_1‡

It follows from the representation $[5.53]$ that

$$g(x_1) = \int_{-\infty}^{\infty} e^{x_1 t_1} \, d\tau(t_1), \qquad \ldots [5.54]$$

where

$$\tau(t_1) = \sum_{p, \, p'=1}^{P} \sigma(t_1, x_2^p + x_2^{p'}) \eta_p \eta_{p'}.$$

Since the integral representation of the form $[5.54]$ with a function of bounded variation $\tau(t_1)$ is unique and since $g(x_1)$ is an exponentially

---

† If we had taken $P = 2$, $x_2^1 = \tfrac{1}{2}(x_2 + h)$, $x_2^2 = \tfrac{1}{2}(x_2 - h)$, $\eta_1 = 1/h$ and $\eta_2 = -1/h$ and had performed the limiting process $h \to 0$ in $[5.52]$ we would have found that the second derivative of $f(x_1, x_2)$ with respect to x_2 also is an exponentially convex function of x_1.

‡ In order to see this, we introduce the continuous hermitian-positive function of y

$$f(iy, x_2) = \int_{-\infty}^{\infty} e^{iyt_1} \, d_{t_1} \sigma(t_1, x_2). \qquad (\alpha)$$

The second derivative of $f(iy, x_2)$ with respect to x_2 is a function of the same type, therefore

$$f_{x_2 x_2}(iy, x_2) = \int_{-\infty}^{\infty} e^{iyt_1} \, d_{t_1} \phi(t_1, x_2),$$

where $\phi(t_1, x_2)$ is a non-decreasing function of t_1 which is normalized in the same way as $\sigma(t_1, x_2)$. It follows from the last representation that

$$f(iy, x_2) = A(y) + x_2 B(y) + \int_0^{x_2} (x_2 - u) \left[\int_{-\infty}^{\infty} e^{iyt_1} \, d_{t_1} \phi(t_1, u) \right] du. \qquad (\beta)$$

Comparing (α) and (β) and performing an inversion (see end of Section **4.1**), we find that

$$\sigma(t_1, x_2) = C(t_1) + x_2 D(t_2) + \int_0^{x_2} (x_2 - u) \phi(t_1, u) \, du,$$

whence it follows not only that the function $\sigma(t_1, x_2)$ is continuous in x_2 but also that it is differentiable.

convex function, it follows that $\tau(t_1)$ is a non-decreasing function. From the fact that $\tau(-\infty) = 0$ it follows that $\tau(t_1) \geqq 0$ i.e. that

$$\sum_{p,\,p'=1}^{P} \sigma(t_1, x_2^p + x_2^{p'})\eta_p\eta_{p'} \geqq 0$$

for any $x_2^p \in (-\frac{1}{2}l_2, \frac{1}{2}l_2)$ and any η_p. Therefore $\sigma(t_1, x_2)$ is an exponentially convex function of t_2 with the parameter t_1. Applying Theorem 5.5.4 again we find that

$$\sigma(t, x_2) = \int_{-\infty}^{\infty} e^{t_2 x_2}\, d_{t_2}\omega(t_1, t_2),$$

where $\omega(t_1, t_2)$ is some non-decreasing bounded function of t_2 for any fixed t_1. Putting this expression into $[5.54]$ we obtain that

$$f(x_1, x_2) = \int_{-\infty}^{\infty} e^{t_1 x_1}\, d_{t_1}\int_{-\infty}^{\infty} e^{t_2 x_2}\, d_{t_2}\omega(t_1, t_2).$$

The transition from this representation to the required one presents no difficulty.

18. Let $I_{a,b}$ denote a certain interval in m-dimensional Euclidean space E_m. Further, let the continuous function $f(x; y)$ ($x \in I_{a,b}$, $y \in E_n$) be given over the topological product $I_{a,b} \times E_n$.

Prove that in order that the function $f(x; y)$ be representable in the form

$$f(x; y) = \int_{E_m \times E_n} e^{(x,\,t)} e^{i(y,\,u)}\, d\omega(t; u),$$

where $\omega(t; u)$ is a non-decreasing point function in $E_m \times E_n$ it is necessary and sufficient that all forms

$$\sum_{\alpha,\,\beta=1}^{N} f(x^\alpha + x^\beta;\ y^\alpha - y^\beta)\xi_\alpha\bar{\xi}_\beta$$

should be non-negative.

Prove also that the stated integral representation is unique.

(A proposition close to this was proved by DEVINATZ[1] with the aid of the spectral theory of operators).

Hint. Show, using the procedure of the previous example that the function

$$F(y) = \sum_{p,\,p'=1}^{P} f(x^p + x^{p'};\ y)\eta_p\eta_{p'}$$

belongs to the class $P(E_m)$ and use the fact that functions of this class permit a known and unique integral representation.

Appendix

STIELTJES CONTINUED FRACTIONS

1. The starting point of Stieltjes' investigation is the infinite continued fraction of the form

$$\cfrac{1}{c_1 z + \cfrac{1}{c_2 + \cfrac{1}{c_3 z + \ddots}}},$$

where $c_1, c_2, c_3, \ldots$ are positive numbers and the variable z can take on arbitrary complex values. The successive even and odd denominators in this fraction have different structure and this has its effect on the structure of the approximating fractions of even and odd order. The results that follow from this fact are made more obvious if, following KREIN[12], one changes the notation† and writes the continued fraction in the form

$$l_0 + \cfrac{1}{m_1 z + \cfrac{1}{l_1 + \cfrac{1}{m_2 z + \ddots}}}, \qquad \ldots [0.1]$$

where $l_0 \geqq 0$ (Stieltjes has $l_0 = 0$) and the remaining $l_k (= c_{2k})$ and all $m_k (= c_{2k-1})$ are positive.

This continued fraction is related to the system of two finite difference equations of first order

$$\begin{aligned} \eta_n - \eta_{n-1} &= l_n \theta_n, \\ \theta_{n+1} - \theta_n &= m_{n+1} z \eta_n \end{aligned} \qquad (n = 1, 2, 3, \ldots). \qquad \ldots [0.2]$$

† The choice of notation finds its justification in the mechanical interpretation, which we shall discuss in the next sub-section.

If we put

$$\eta_0 = 1, \quad \theta_1 = m_1 z,$$

the polynomials η_n $(n = 0, 1, 2, ...)$ will be the denominators $U_n(z)$ of the approximating fractions

$$\frac{V_n(z)}{U_n(z)}$$

of odd order and the polynomials θ_n $(n = 1, 2, 3, ...)$ the denominators $G_n(z)$ of the approximating fractions

$$\frac{H_n(z)}{G_n(z)}$$

of even order. The corresponding numerators can be obtained by putting

$$\eta_0 = l_0, \quad \theta_1 = 1 + l_0 m_1 z.$$

The system of equations [0.2] is equivalent to the single equation

$$b_n y_{n+1} + a_n y_n + b_{n-1} y_{n-1} = \lambda y_n \quad (n = 1, 2, 3, ...),$$

where

$$b_n = \frac{1}{l_{n+1}\sqrt{m_{n+1}m_{n+2}}} \quad (n = 0, 1, 2, ...), \qquad ...[0.3]$$

$$a_n = \frac{1}{m_{n+1}}\left\{\frac{1}{l_n} + \frac{1}{l_{n+1}}\right\} \quad (n = 1, 2, 3, ...), \qquad ...[0.4]$$

$$y_n = (-1)^n \sqrt{m_{n+1}}\, \eta_n \quad (n = 0, 1, 2, ...),$$

$$\lambda = -z.$$

If we also put

$$a_0 = \frac{1}{m_1}\frac{1}{l_1},$$

we are led to a certain $\mathscr{J}$-matrix and therefore to polynomials $P_k(\lambda)$ and $Q_k(\lambda)$ and a certain positive sequence $\{s_k\}_0^\infty$. Here however something specific emerges.

Indeed, we can write down the relation

$$\sum_{i=0}^{n} a_i \xi_i^2 - 2\sum_{i=0}^{n-1} b_i \xi_i \xi_{i+1}$$

$$= \frac{1}{l_{n+1}m_{n+1}}\xi_n^2 + \sum_{i=0}^{n-1}\frac{1}{l_{i+1}}\left\{\frac{1}{\sqrt{m_{i+1}}}\xi_i - \frac{1}{\sqrt{m_{i+2}}}\xi_{i+1}\right\}^2 \geqq 0.$$

Therefore remembering equation [*1.10*], which has the form

$$\sum_{i,\,k=0}^{n} s_{i+k+1} x_i x_k = \sum_{i=0}^{n} a_i \xi_i^2 - 2 \sum_{i=0}^{n-1} b_i \xi_i \xi_{i+1}, \qquad \ldots[0.5]$$

we reach the conclusion that in the present case the sequence $\{s_k\}_0^\infty$ is positive not only relative to the whole axis but also relative to the semi-axis $x \geqq 0$.

This result permits an inversion, namely: *if a sequence $\{s_k\}_0^\infty$ is positive relative to the semi-axis $x \geqq 0$, one can find a Stieltjes continued fraction [0.1] to which this positive sequence belongs.*

Proof. We construct the $\mathscr{J}$-matrix for the sequence $\{s_k\}_0^\infty$. On the basis of equation [0.5] we find that

$$\sum_{i=0}^{n} a_i \xi_i^2 - 2 \sum_{i=0}^{n-1} b_i \xi_i \xi_{i+1} \geqq 0. \qquad \ldots[0.6]$$

Then we take an arbitrary $m_1 > 0$ and define l_1 by the formula

$$a_0 = \frac{1}{m_1} \frac{1}{l_1},$$

so that $l_1 > 0$. Equation [0.3] for $n = 0$, which is

$$b_0 = \frac{1}{l_1 \sqrt{m_1 m_2}}$$

allows us to find a positive m_2 and then equation [0.4] for $n = 1$,

$$a_1 = \frac{1}{m_2} \left\{ \frac{1}{l_1} + \frac{1}{l_2} \right\},$$

serves to determine l_2. We must only convince ourselves that $l_2 > 0$. For this purpose it is sufficient to take inequality [0.6] and insert $n = 1$, $\xi_i = \sqrt{m_{i+1}}$ ($i = 0, 1$). The continuation of this procedure is obvious and leads to all the elements of the Stieltjes continued fraction.

2. The system of finite difference equations [0.2] allows a useful mechanical interpretation, which was the starting point for the far-reaching constructions due to Krein relating to an arbitrary string.

Let us imagine a weightless, absolutely flexible thread stretched under the action of a unit force between the points $x = 0$ and $x = L$ ($\leqq \infty$) and carrying point masses whose weights and abscissae are

$$m_1, \xi_1; \quad m_2, \xi_2; \quad m_3, \xi_3; \quad \ldots,$$

with

$$\xi_1 = l_0, \quad \xi_2 - \xi_1 = l_1, \quad \xi_3 - \xi_2 = l_2, \ldots$$

We shall assume that the string performs small harmonic oscillations in which a point ξ moves parallel to the ordinate, according to the law

$$u(\xi, t) = \eta(\xi, \omega) \sin \omega t,$$

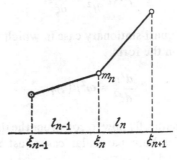

FIG. 3

where the frequency ω is the same for all points and $\eta(\xi, \omega)$ represents the so-called amplitude function. For the amplitude functions of concentrated masses we introduce the following notation

$$\eta_n = \eta(\xi_n, \omega).$$

For an arbitrary instant of time t we write down the equations of motion for the mass m_n using D'Alembert's principle. In doing this we bear in mind the smallness of the oscillations, which permits us to replace the sines of the angles of inclination of the string segments to the x-axis by their tangents. These tangents are

$$\theta_n \sin \omega t = \frac{\eta_{n+1} - \eta_n}{l_n} \sin \omega t. \qquad \ldots[0.7]$$

Therefore the equations of motion take the form

$$m_n \eta_n \omega^2 \sin \omega t + \theta_n \sin \omega t - \theta_{n-1} \sin \omega t = 0$$

or

$$\theta_n - \theta_{n-1} = -m_n \eta_n \omega^2. \qquad \ldots[0.8a]$$

On the other hand, it follows from $[0.7]$ that

$$\eta_{n+1} - \eta_n = l_n \theta_n. \qquad \ldots[0.8b]$$

The equations $[0.8a]$ and $[0.8b]$ are the same as equations $[0.2]$ if one puts $z = -\omega^2$.

Side by side with a string consisting of a finite number of masses fixed on a thread, a string with a continuous mass distribution has also been well studied, or more precisely a string possessing a finite density $\mu(\xi) \geq 0$ at every point. The equations describing the free oscillations of such a string have the form

$$\mu(\xi) \frac{\partial^2 u}{\partial t^2} = \frac{\partial^2 u}{\partial \xi^2}.$$

If one considers the quasi-stationary case in which $u(\xi, t) = \eta(\xi) \sin \omega t$, the equation takes on the form

$$\frac{d^2 \eta}{d\xi^2} = \omega^2 \mu(\xi)\eta.$$

The question arises naturally how to construct the theory of a " general " string, which includes both particular cases just mentioned. Such a theory was constructed by KREIN[11].

3. We give a series of relations (identities and inequalities) which will be needed in the following work. We shall confine ourselves to proof of the inequalities; the verification of the identities is left to the reader. We note that everywhere $\lambda = -z$.

$$U_n(0) = 1, \quad V_n(0) = l_0 + l_1 + \dots + l_n \quad (n = 0, 1, 2, \dots),$$

$$\frac{V_n\ z)}{U_n(z)} = l_0 - \frac{Q_n(\lambda)}{P_n(\lambda)} \quad (n = 1, 2, 3, \dots),$$

$$U_n(z) = m_1 \dots m_n l_1 \dots l_n z^n + \dots,$$

$$U_n(z) = (-1)^n \frac{1}{\sqrt{m_{n+1}}} P_n(\lambda),$$

$$P_n(0) = (-1)^n \sqrt{m_{n+1}},$$

$$Q_n(0) = (-1)^{n-1}(l_1 + l_2 + \dots + l_n)\sqrt{m_{n+1}}. \quad \dots[0.9]$$

The polynomial $G_n(z)$ is fully characterized by the following properties:

$$\mathfrak{S}\{G_n(-\lambda)\lambda^k\} = 0 \quad (k = 0, 1, \dots, n-2),$$

$$G_n(0) = 0, \ G_n(z) = m_1 \dots m_n l_1 \dots l_{n-1} z^n + \dots$$

Further

$$\frac{H_{n+1}(z)}{G_{n+1}(z)} - \frac{V_n(z)}{U_n(z)} = \frac{1}{U_n(z)G_{n+1}(z)}. \quad \dots[0.10]$$

For $z > 0$ the following inequalities hold

$$\left.\begin{array}{l} G_n(z) > z(m_1 + m_2 + \ldots + m_n), \\ U_n(z) > m_1 z(l_1 + l_2 + \ldots + l_n), \end{array}\right\} \qquad \ldots[0.11]$$

$$G_{n+1}(z)U_n(z) < \tfrac{1}{2}\{G_{n+1}(z) + U_n(z)\}^2$$

$$< \tfrac{1}{2} \exp\left(2\sum_1^n l_k\right) \cdot \exp\left(2z \sum_1^{n+1} m_k\right), \qquad \ldots[0.12]$$

$$\frac{V_n(z)}{U_n(z)} < \frac{V_{n+1} z)}{U_{n+1}(z)} < \frac{H_{n+1}(z)}{G_{n+1}(z)} < \frac{H_n(z)}{G_n(z)}. \qquad \ldots[0.13]$$

The inequalities $[0.11]$ follow directly from the definitions

$$U_n(z) - U_{n-1}(z) = l_n G_n(z),$$

$$G_{n+1}(z) - G_n(z) = m_{n+1} z U_n(z),$$

$$U_0(z) = 1, \quad G_1(z) = m_1 z.$$

The same definitions give us

$$G_n(z) + U_n(z) \leqq (1 + l_n)[G_n(z) + U_{n-1}(z)]$$

and

$$G_{n+1}(z) + U_n(z) \leqq (1 + m_{n+1} z)[G_n(z) + U_n(z)],$$

whence the inequality

$$G_{n+1}(z) + U_n(z) \leqq (1 + l_n) \ldots (1 + l_1)(1 + m_{n+1} z) \ldots (1 + m_1 z)$$

$$< \exp\left(\sum_1^n l_k\right) \exp\left(z \sum_1^{n+1} m_k\right)$$

follows and with it also inequality $[0.12]$.

As for inequality $[0.13]$ it is a consequence of the fact that the number z is positive and that all the parameters of the continued fraction $[0.1]$ are also positive.

4. THEOREM 0.4 (STIELTJES[3]). *For the Stieltjes moment problem*

$$s_k = \int_0^\infty u^k \, d\sigma(u) \quad (k = 0, 1, 2, \ldots), \qquad \ldots[0.14]$$

to be determinate when the sequence $\{s_k\}_0^\infty$ belongs to a certain continued fraction $[0.1]$ (i.e. is positive relative to the semi-axis $x \geqq 0$), it is necessary and sufficient that at least one of the following series be divergent

$$\sum_1^\infty m_k, \quad \sum_1^\infty l_k. \qquad \ldots[0.15]$$

Proof. We take an arbitrary number $z = -\lambda > 0$ and write down the inequalities

$$I_1 = \int_0^\infty \left[\frac{U_n(-u)}{U_n(-\lambda)}\right]^2 \frac{d\sigma(u)}{u-\lambda} \geqq 0,$$

$$I_2 = \int_0^\infty \frac{\lambda}{u}\left[\frac{G_{n+1}(-u)}{G_{n+1}(-\lambda)}\right]^2 \frac{d\sigma(u)}{u-\lambda} \leqq 0,$$

where $\sigma(u)$ is any solution of the moment problem [0.14].
Since

$$I_1 = \int_0^\infty \frac{P_n(u)}{P_n(\lambda)}\{1 + A_1(u-\lambda) + \ldots + A_n(u-\lambda)^n\}\frac{d\sigma(u)}{u-\lambda}$$

$$= \int_0^\infty \frac{P_n(u)}{P_n(\lambda)}\frac{d\sigma(u)}{u-\lambda} = \int_0^\infty \frac{d\sigma(u)}{u-\lambda} + \frac{Q_n(\lambda)}{P_n(\lambda)}$$

$$= \int_0^\infty \frac{d\sigma(u)}{u+z} + l_0 - \frac{V_n(z)}{U_n(z)},$$

we have that

$$\frac{V_n(z)}{U_n(z)} \leqq l_0 + \int_0^\infty \frac{d\sigma(u)}{u+z}.$$

In a similar manner we find from the inequality $I_2 \leqq 0$ that

$$l_0 + \int_0^\infty \frac{d\sigma(u)}{u+z} \leqq \frac{H_{n+1}(z)}{G_{n+1}(z)}.$$

Using this inequality we find that for any two solutions $\sigma_1(u)$ and $\sigma_2(u)$ of this moment problem we have

$$\left|\int_0^\infty \frac{d\sigma_1(u)}{u+z} - \int_0^\infty \frac{d\sigma_2(u)}{u+z}\right| \leqq \frac{H_{n+1}(z)}{G_{n+1}(z)} - \frac{V_n(z)}{U_n(z)}.$$

Hence, by virtue of the identity [0.10] and the inequality [0.11] it follows that

$$\left|\int_0^\infty \frac{d\sigma_1(u)}{u+z} - \int_0^\infty \frac{d\sigma_2(u)}{u+z}\right| \leqq \frac{1}{m_1 z^2}\frac{1}{\sum_1^n l_k \cdot \sum_1^{n+1} m_k}.$$

Therefore, if at least one of the series [0.15] is divergent, the analytic functions

$$\int_0^\infty \frac{d\sigma_1(u)}{u+z}, \quad \int_0^\infty \frac{d\sigma_2(u)}{u+z},$$

which are regular in the z-plane cut along the negative real axis, coincide on the positive half of this axis. Therefore these functions are identical and so $\sigma_1(u)$ and $\sigma_2(u)$ are not different solutions of the moment problem.

Thus, sufficiency has been proved.

In order to prove necessity we write down two representations

$$\frac{V_n(z)}{U_n(z)} = l_0 + \int_0^\infty \frac{d\tau_n(u)}{u+z}, \quad \frac{H_{n+1}(z)}{G_{n+1}(z)} = l_0 + \int_0^\infty \frac{d\omega_n(u)}{u+z},$$

where $\tau_n(u)$ and $\omega_n(u)$ are sectionally constant non-decreasing functions each of which is a solution of the truncated problem

$$s_k = \int_0^\infty u^k \, d\sigma(u) \quad (k = 0, 1, \ldots, m). \qquad \ldots [0.14a]$$

The correctness of these representations follows from the general constructions of Section **1.1** in Chapter 2 if one notes that all zeros of the polynomials $U_n(-u)$ and $G_{n+1}(-u)$ lie on the semi-axis $u \geqq 0$. By virtue of the relations $[0.10]$ and $[0.12]$ we can write down for $z > 0$ the inequality

$$\int_0^\infty \frac{d\omega_n(u)}{u+z} - \int_0^\infty \frac{d\tau_n(u)}{u+z} = \frac{H_{n+1}(z)}{G_{n+1}(z)} - \frac{V_n(z)}{U_n(z)}$$

$$> 2 \exp\left(-2 \sum_1^n l_k\right) \exp\left(-2z \sum_1^{n+1} m_k\right).$$

If we now assume that both series $[0.15]$ converge, then, using Helly's theorem, we can obtain two solutions $\omega(u)$ and $\tau(u)$ of the problem $[0.14]$ for which we have, for any finite $z > 0$,

$$\int_0^\infty \frac{d\omega(u)}{u+z} - \int_0^\infty \frac{d\tau(u)}{u+z} \geqq 2 \exp\left(-2 \sum_1^\infty l_k\right) \exp\left(-2z \sum_1^\infty m_k\right) > 0,$$

and this is impossible if the moment problem $[0.14]$ is determinate.

Thus the necessity proof is also complete.

5. Assume that a sequence $\{s_k\}_0^\infty$ is given, which is positive relative to the semi-axis. In this case the moment problem $[0.14]$ has a solution $\sigma(u)$ and each such solution can be considered as a solution of the Hamburger moment problem

$$s_k = \int_{-\infty}^\infty u^k \, d\tau(u) \quad (k = 0, 1, 2, \ldots). \qquad \ldots [0.16]$$

The question arises naturally whether the moment problem $[0.16]$ has solutions which are not solutions of the moment problem $[0.14]$, or in other words solutions which do not maintain a constant value for $u < 0$. In order to answer this question one must obtain a criterion for the determinateness of problem $[0.16]$ expressed in terms of the Stieltjes parameters l_k and m_k. Hamburger found such a criterion.

THEOREM **0.5** (HAMBURGER[3]). *If the sequence $\{s_k\}_0^\infty$ belongs to the continued fraction $[0.1]$ then it is necessary and sufficient for the moment problem $[0.16]$ to be determinate that*

$$\sum_{n=1}^{\infty} m_{n+1}(l_1 + l_2 + \ldots + l_n)^2 = \infty. \qquad \ldots[0.17]$$

Comparing this theorem with Theorem **0.4** we come to the conclusion that the Stieltjes moment problem $[0.14]$ is determinate and the Hamburger moment problem $[0.16]$ is indeterminate if and only if

$$\sum_{n=1}^{\infty} l_n = \infty, \quad \sum_{n=1}^{\infty} m_{n+1}(l_1 + l_2 + \ldots + l_n)^2 < \infty.$$

Proof. In the Addenda and Problems to Chapter 2 (see example **10**) it has been shown that for the Hamburger moment problem to be determinate it is necessary and sufficient that at least one of the series

$$\sum_{0}^{\infty} [P_k(0)]^2, \quad \sum_{1}^{\infty} [Q_k(0)]^2.$$

should diverge. In the case under consideration equation $[0.9]$ ensures that these series have the form

$$\sum_{k=1}^{\infty} m_k, \quad \sum_{n=1}^{\infty} m_{n+1}(l_1 + l_2 + \ldots + l_n)^2$$

and the correctness of the theorem follows from the fact that the first series cannot diverge if the second converges.

6. The Stieltjes moment problem can be generalized as follows: *given the infinite sequence $\{s_k\}_0^\infty$ and the point set E formed by the intervals $(-\infty, \alpha]$ and $[\beta, \infty)$ where $\beta > \alpha$ it is required to find a non-decreasing function $\sigma(u)$ such that*

$$s_k = \int_{-\infty}^{\infty} u^k \, d\sigma(u) \quad (k = 0, 1, 2, \ldots). \qquad \ldots[0.18]$$

This problem was the subject of the dissertation by SHVETSOV[1], work done under the supervision of the late A. M. Danilevskii. The basic results of Shvetsov's paper are the following:

a) For the moment problem [*0.18*] to be soluble it is necessary and sufficient that the following two moment problems be soluble:

$$s_k = \int_{-\infty}^{\infty} u^k \, d\sigma(u), \quad s_k^* = \int_{-\infty}^{\infty} u^k \, d\sigma^*(u) \quad (k = 0, 1, 2, \ldots), \quad \ldots [0.19]$$

where

$$s_k^* = s_{k+2} - (\alpha + \beta)s_{k+1} + \alpha\beta s_k.$$

b) Let both Hamburger problems [*0.19*] be soluble. For any positive integer m and any non-real z circles $K_m(z)$ and $K_{m-1}^*(z)$ in the planes of w and w^* belong to these problems. We denote by $K_{m-1}(z; \alpha, \beta)$ the circle in the w plane into which $K_{m-1}^*(z)$ is transformed in the mapping:

$$w = \frac{w^* - (z - \alpha - \beta)s_0 - s_1}{(z - \alpha)(z - \beta)}$$

The circles $K_m(z)$ and $K_{m-1}(z; \alpha, \beta)$ intersect, forming a crescent $L_m(z)$. The vertices of this crescent are the points

$$A_m = \frac{Q_{m+1}(z)P_m(\alpha) - Q_m(z)P_{m+1}(\alpha)}{P_{m+1}(z)P_m(\alpha) - P_m(z)P_{m+1}(\alpha)},$$

$$B_m = \frac{Q_{m+1}(z)P_m(\beta) - Q_m(z)P_{m+1}(\beta)}{P_{m+1}(z)P_m(\beta) - P_m(z)P_{m+1}(\beta)}.$$

The angle of the crescent $L_m(z)$ is

$$\arg \frac{\beta - z}{z - \alpha}.$$

The crescent $L_{m+1}(z)$ lies inside the crescent $L_m(z)$. As the point z approaches a point x in the interval (α, β) the crescent $L_m(z)$ degenerates into a segment of the real axis the length of which is

$$\delta_m(x) = \frac{\beta - \alpha}{(\beta - x)(x - \alpha)} \left| \frac{\sum_{0}^{m} P_k(\alpha)P_k(\beta)}{\left\{ \sum_{0}^{m} P_k(\alpha)P_k(x) \right\}\left\{ \sum_{0}^{m} P_k(\beta)P_k(x) \right\}} \right|.$$

c) For the moment problem [*0.18*] to be determinate it is necessary and sufficient that at least one of the two Hamburger moment problems [*0.19*] should be determinate.

d) For the moment problem [*0.18*] to be determinate it is necessary that

$$\lim_{m \to \infty} \delta_m(x) = 0$$

for every x in the interval (α, β) and sufficient that this equation should hold for at least one such x.

BIBLIOGRAPHY

AKHIEZER, N. I.
1. 1941. Infinite Jacobi matrices and the moment Problem. *Uspekhi Matem. Nauk*, 9 (in Russian).
2. 1945. On a proposition of Kolmogorov and a suggestion of Krein. *Dokl. Akad. Nauk SSSR*, **50** (in Russian).
3. 1956. *Theory of approximation* Frederick Ungar, New York (in English).
4. 1949. Markov's moment problem for an arbitrary number of intervals *Ukrain, matem. Zhur.* No. 3, pp. 41–50 (in Russian).
5. 1950. On solutions of the power moment problem in the indeterminate case. *Commun. Soc. math. Kharkov*, **20** (in Russian).
6. 1954. On a generalization of the Fourier Transformation and the Paley-Wiener theorem. *Dokl. Akad. Nauk SSSR*, **96**, 889–892 (in Russian).
7. 1956. On the weighted approximation of continuous functions by polynomials on the whole number axis. *Uspekhi Matem. Nauk.* (in Russian).

AKHIEZER, N. I. & GLAZMAN, I. M.
1. 1961. *Theory of linear operators in Hilbert Space* Frederick Ungar, New York (2 vols.).
2. 1957. On some classes of continuous functions that generate hermitian-positive kernels. *Commun. Soc. math. Kharkov*, **25** (in Russian).

AKHIEZER, N. I. & KREIN, M. G.
1. 1962. *Some questions in the theory of moments* Amer. Math. Soc. New York.
2. 1952. On a generalization of the lemma of Schwartz and Löwner. *Commun. Soc. math. Kharkov*, **23** (in Russian).

BANACH, S.
1. 1948. *A course of functional analysis* (in Ukrainian). Radyanska Shkola, Kiev.

BEREZANSKII, YU. M.
1. 1956. Expansion in terms of characteristic functions of a second order partial differential equation. *Trud. Mosk. matem. ob.* 5 (in Russian).

BERNSTEIN, S. N.
1. 1926. Leçons sur les propriétés extrémales des fonctions analytiques d'une variable réelle. Collection Borel. Gauthier-Villars, Paris (in French).
2. 1929. Sur les fonctions absolument monotones. *Acta Math.* **52**, 1–66 (in French).
2. 1929. Sur les fonctions absolument monotones. *Acta Math.* **52**, 1–66.

BOCHNER, S.
1. 1932. *Vorlesung über Fouriersche Integrale* Leipzig Verlag. English translation: Lectures on Fourier Integrals, *Ann. Math. Stud.*, **42**, Princeton Univ. Press.
2. 1933. Monotone Funktionen, Stieltjessche Integrale und harmonische Funktionen. *Math. Ann.* **108**, 378–410 (in German).
3. 1955. *Harmonic Analysis and the Theory of Probability* Univ. Calif. Press.

CARATHÉODORY, C.

1. 1911. Über den Variabilitätsbereich der Fourierschen Konstanten von positiven harmonischen Funktionen. *Rend. Circ. mat. Palermo* **32**, 193–217.

CARLEMAN, T.

1. 1922. Sur le problème des moments. *C.R. Acad. Sci. Paris* **174**, 1680–1682.
2. 1923. *Sur les équations intégrales singulières à noyau réel et symmétrique. Uppsala Univ. Årsskr.*
3. 1926. *Les fonctions quasi-analytiques.* Collection Borel. Gauthier-Villars, Paris.

CHEBYSHEV, P. L.

1961. *Oeuvres.* Chelsea Pub. Co., New York (2 vols.).
1. On limiting values of integrals.
2. On the representation of limiting values of integrals in terms of residues of integrals.
3. On residues of integrals giving approximate values of integrals.
4. Two theorems on probabilities.
5. On the expansion into a continued fraction of series in non-converging powers of the variable.

COTLAR, M.

1. 1954. *El problema de los momentos y la teoria de operatores hermitianos.* Montevideo. (UNESCO publication).

CRAMÉR, H. & WOLD, H.

1. 1936. Some theorems on distribution functions. *J. L. M. S.* **11**, 290–294.

DENNIS, J. J. & WALL, H. S.

1. 1945. The limit-circle case for a positive-definite *J*-fraction. *Duke Math. J.* **12**, 255–273.

DEVINATZ, A.

1. 1953. Integral representations of positive-definite functions. *Trans. Am. Math. Soc.* **74**, 56–77.

FISCHER, E.

1. 1911. Über das Carathéodorysche Problem, Potenzreihen mit positivem reellen Teil betreffend. *Rend. Circ. mat. Palermo* **32**, 240–256.

GEL'FAND, I. M. & LEVITAN, B. M.

1. 1951. On the determination of a differential equation from its spectral function. *Izv. Akad. Nauk SSSR, ser. matem.* **15** (in Russian).

GERONIMUS, YA. L.

1. 1948. Polynomials orthogonal on a circle and their applications. *Commun. Soc. mat. Kharkov* **19**, 35–120 (in Russian).
2. 1948. On some asymptotic properties of polynomials. *Matem. Sbornik* **23**, 77–88 (in Russian).
3. 1950. On asymptotic properties of polynomials that are orthogonal on the unit circle and on some properties of positive-harmonic functions. *Izv. Akad. Nauk SSSR, ser. matem.* **14** (in Russian).
4. 1960. *Polynomials orthogonal on a circle and interval.* Pergamon Press, Oxford (in English).

GLAZMAN, I. M.
1. 1950. On a class of solutions of the classical moment problem. *Commun. Soc. math. Kharkov* **20** (in Russian).

GLAZMAN, I. M. & NAIMAN, P. B.
1. 1955. On the convex hull of orthogonal spectral functions. *Dokl. Akad. Nauk SSSR* **102** (in Russian).

GNEDENKO, B. V.
1. 1937. On characteristic functions. *Bull. Mosc. Univ.* (in Russian).
2. 1962. *The theory of probability.* Chelsea Pub. Co., New York.

GOKHMAN, E. KH.
1. 1958. *The Stieltjes Integral and its applications.* Moscow (in Russian).

GURARII, V. P.
1. 1960. Generalization of the Fourier transformation and the Paley-Wiener theorem. *Dokl. Akad. Nauk SSSR* **130**, 959–962. (English translation: *Soviet Math.* **1** (1960) 102–105).

HAMBURGER, H.
1. 1919. Beiträge zur Konvergenztheorie der Stieltjesschen Kettenbrüche. *Math. Z.* **4**, 186–222.
2. 1920. Über die Konvergenz eines mit einer Potenzreihe assoziierten Kettenbruchs. *Math. Ann.* **81**, 31–45.
3. 1920, 1921. Über eine Erweiterung des Stieltjesschen Momentenproblems. *Math. Ann.* **81**, 235–319, **82**, 120–164; **82**, 168–187.
4. 1944. Hermitian transformations of deficiency index (1, 1), Jacobi matrices and undetermined moment problems. *Amer. J. Math.* **66**, 489–522.
5. 1944. Contributions to the theory of closed Hermitian transformations of deficiency index (*m, m*). *Ann. of Math.* **45**, 59–99.

HARDY, G. H.
1. 1917. On Stieltjes' " problème des moments ". *Messenger of Math.* **46**, 175–182 & **47**, 81–88.
2. 1949. *Divergent series.* O.U.P., London.

HARTMAN, P.
1. 1948. Differential equations with non-oscillatory eigenfunctions. *Duke Math. J.* **15**, 697–709.

HAUSDORFF, F.
1. 1921. Summationsmethoden und Momentenfolgen. *Math. Z.* **9**, I, 74–109, II, 280–299.
2. 1923. Momentenprobleme für ein endliches Intervall. *Math. Z.* **16**, 220–248.

HAVILAND, E. K.
1. 1934. On the theory of absolutely additive distribution functions. *Amer. J. Math.* **56**, 625–658.
2. 1935. On the moment problem for distribution functions in more than one dimension, I. *Amer. J. Math.* **57**, 562–568.
3. 1936. On the moment problem for distribution functions in more than one dimension, II. *Amer. J. Math.* **58**, 164–168.

HELLINGER, E.
1. 1922. Zur Stieltjesschen Kettenbruchtheorie. *Math. Ann.* **86**, 18–29.

HERGLOTZ, G.
1. 1911. Über Potenzreihen mit positivem reellen Teil im Einheitskreis. *Ber. Ver. Sächs. Ges. d. Wiss. Leipzig.* **63**, 501–511.

HILDEBRANDT, T. & SCHOENBERG, I. J.
1. 1933. On linear functional operators and the moment problem for a finite interval in one or several dimensions. *Ann. of Math.* (*Ser.* 2) **34**, 317–328.

HIRSCHMAN, I. I. & WIDDER, D. V.
1. 1955. *The Convolution-transform.* Princeton Univ. Press.

KANTOROVICH, L. V.
1. 1937. On the moment problem for a finite interval. *Dokl. Akad. Nauk SSSR* **14**, 531–537. (In English.)

KATS, I.
1. 1956. On integral representations of analytic functions which map the upper half-plane on part of itself. *Uspekhi matem. Nauk.* (in Russian).

KHINCHIN, A. YA.
1. 1938. *Limit theorems for sums of random variables.* O.N.T.I., Moscow (in Russian).

KILPI, Y.
1. 1957. Über das komplexe Momentenproblem. *Ann. Acad. Sc. Fenn.* A **236** (32 p.).

KOLMOGOROV, A. N.
1. 1941. Stationary sequences in Hilbert Space. *Bull. Mosc. Univ.* (1) (in Russian).
2. 1941. Interpolation and extrapolation of stationary random sequences. *Izv. Akad. Nauk SSSR ser. matem.* **5**, 11–14. (In German.)

KORÀNYI, A.
1. 1956. On a theorem of Löwner and its connections with resolvents of self-adjoint transformations. *Acta Sci. Math. Szeg.* **17**, 63–70 (in English).

KRASNOSEL'SKII, M. A. & KREIN, M. G.
1. 1947. Basic theorems on the extension of hermitian operators and some applications of these to the theory of orthogonal polynomials and to the moment problem. *Uspekhi mat. Nauk.* (in Russian).

KREIN, M. G.
1. 1940. On the problem of continuation of hermitian-positive continuous functions. *Dokl. Akad. Nauk SSSR* **26**, 17–22. (In French.)
2. 1944. On hermitian operators with unit deficiency indices. *Dokl. Akad. Nauk SSSR* **43**, 323–326. (In English.)
3. 1944. On the logarithm of a hermitian-positive function which permits unlimited expansion. *Dokl. Akad. Nauk SSSR* **45**, 91–94. (In English.)
4. 1945. On a generalization of an investigation by Szegö, Smirnov and Kolmogorov. *Dokl. Akad. Nauk SSSR* **46**, 91–94. (In English.)
5. 1945. On an extrapolation problem due to Kolmogorov. *Dokl. Akad. Nauk SSSR* **46**, 306–309. (In English.)

6. 1946. On a general method for expanding positive-definite kernels into elementary products. *Dokl. Akad. Nauk SSSR* **53**, 3–6. (In English.)
7. 1947. On the theory of entire functions of exponential type. *Izv. Akad. Nauk SSSR. ser. matem.* **11**, 309 (in Russian with Summary in English).
8. 1948. On hermitian operators with directing functionals. *Zbir. prats inst. matem. Akad. Nauk URSR* (in Ukrainian).
9. 1951. Chebyshev's and Markov's ideas in the theory of limiting values of integrals and their further development, (with editorial assistance of P. G. Rekhtman) *Uspekhi matem. Nauk.* **44** (in Russian).
10. 1952. On the indeterminate case in the boundary value problem for a Sturm-Liouville equation in the interval (0, ∞). *Izv. Akad. Nauk SSSR ser. matem.* **16** (in Russian).
11. 1952. On a generalization of an investigation by Stieltjes. *Dokl. Akad. Nauk SSSR* **87** (in Russian).
12. 1952. Some new problems in the theory of oscillations of Sturm systems. *Priklad. matem. i mekh.* **16**, 555–568 (in Russian).
13. 1953. On the transition function for a homogeneous second order boundary value problem. *Dokl. Akad. Nauk SSSR* **88**, 405–408 (in Russian).
14. 1955. Continuous analogues of propositions on orthogonal polynomials on the unit circle. *Dokl. Akad. Nauk SSSR* **105**, 637–640 (in Russian).

KREIN, M. G. & REKHTMAN, P. G.

1. 1938. On the Nevanlinna-Pick problem. *Trudy Odess. Gos. Univ.* (*Travaux de l'Univ. d'Odessa*) **2**, 63–68 (in Ukrainian).

LÉVY, P.

1. 1934. Sur les intégrales dont les éléments sont des variables aléatoires indépendentes. *Annali Scu. norm. sup., Pisa* (2), **3**, 337–366 (in French).

LEVIN, B. YA.

1. 1956. *The distribution of zeros of entire functions.* Gosudarstv. Izdat. Tehn.-Teor. Lit. 3 Moscow (in Russian).

LEVITAN, B. M.

1. 1950. *Expansion in terms of characteristic functions of second order differential equations.* Dokl. Akad. Nauk. SSSR **71** (in Russian).

LIVSHITS, M. S.

1. 1944. An application of the theory of hermitian operators to the generalized moment problem. *Dokl. Akad. Nauk SSSR* **44**, 3–7. (In English.)

MARCHENKO, V. A.

1. 1952, 1953. Some problems in the theory of homogeneous linear second order differential operators. *Trudy Mosc. math. soc.* **1, 2** (in Russian).

MARKOV, A. A.

1. 1884. *On some applications of algebraic continued fractions.* (Dissertation) St. Petersburg.
2. 1895. On limiting values of integrals. (Session of the physico-mathematical section of the Academy of Sciences, Feb. 6th. The article is included as a commentary in vol. III of Chebyshev's collected works, cf. CHEBYSHEV 1961.)

3. 1948. Proof of some inequalities due to Chebyshev. *Selected papers.* Moscow.
4. 1948. Extract from a letter to Hermite. *ibid.*
5. 1948. On functions obtained in transforming series into continued fractions. *ibid.*
6. 1948. Two proofs of convergence for certain continued fractions. *ibid.*
7. 1948. New applications of continued fractions. *ibid.*
8. 1948. On limiting values of integrals in connection with interpolation. *ibid.*
9. 1948. On the roots of the equation $e^{x^2} d^m e^{-x^2}/dx^m = 0$. *ibid.*
10. 1924. *Evaluation of probabilities.* Moscow.

MATHIAS, M.

1. 1923. Über positive Fourier-Integrale. *Math. Z.* **16**, 103–125.

MATSAEV, V. I.

1. 1960. On entire functions that have a lower bound. *Dokl. Akad. Nauk SSSR* **132**, 283–286. (English translation: *Soviet Math.* **1** (1960), 548–552.)

NAGY, B. SZ. & KORÀNYI, A.

1. 1957. Relations d'un problème de Nevanlinna et Pick avec la théorie des opérations de l'espace hilbertien. *Acta Math. Acad. Sci. Hung.* **7**, 295–302.
2. 1958. Operatortheoretische Behandlung und Verallgemeinerung eines Problemkreises in der komplexen Funktionentheorie. *Acta Math.* **100**, 171–202.

NAIMARK, M. A.

1. 1940. On self adjoint extensions of the second kind of a symmetric operator. *Izv. Akad. Nauk. SSSR* (ser. matem.) **4**, 90–104. (In English.)
2. 1940. Spectral functions of a symmetric operator. *Izv. Akad. Nauk SSSR ser. matem.* **4**, 53–89 (in Russian, with English summary).
3. 1947. Extremal spectral functions of a symmetric operator. *Izv. Akad. Nauk SSSR ser. matem.* **11**; *Dokl. Akad. Nauk SSSR* **54** (1946), 7–9 (in Russian, with English summary).

NEVANLINNA, R.

1. 1922. Asymptotische Entwickelungen beschränkter Funktionen und das Stieltjessche Momentenproblem. *Ann. Acad. Sci. Fenn.* A **18**, no. 5 (52 pp.).
2. 1929. Über beschränkte analytische Funktionen. *Ann. Acad. Sci. Fenn.* A **32**, no. 7 (75 pp.).

OSTROVSKII, I. V.

1. 1957. Generalization of a theorem due to Krein. *Dokl. Akad. Nauk SSSR* **116**, 742–745 (in Russian).

PERRON, O.

1. 1929. *Die Lehre von den Kettenbrüchen*, 2 Aufl. Teubner, Leipzig.

PICK, G.

1. 1916. Über die Beschränkungen analytischer Funktionen, welche durch vorgegebene Funktionswerte bewirkt sind. *Math. Ann.* **77**, 7–23.
2. 1920. Über beschränkte Funktionen mit vorgegeben Wertzuordnungen. *Ann. Acad. Sci. Fenn.* A **15**, no. 3 (17 pp.).

POLYA, G.

1. 1954. *Mathematics and Plausible Reasoning*. In two vols. I. Induction and Analogy in Mathematics, II. Patterns of Plausible Inference. O.U.P., London.

POLYA, G. & SZEGÖ, G.

1. 1954. *Aufgaben und Lehrsätze aus der Analysis* Vol. **2**, 82. Springer-Verlag Berlin.

POSSE, K.

1. 1886. *Sur quelques applications des fractions continues algébriques.* St. Petersburg.

RADON, J.

1. 1913. Theorie und Anwedungen der absolut additiven Mengenfunktionen. *Sitzb. der Akad. zu Wien* **122**, 1295–1438.

RIESZ, F.

1. 1911. Sur certains systèmes singuliers d'équations intégrales. *Ann. Éc. Norm.* **28**, 33–62.
2. 1913. *Les systèmes d'équations linéaires à une infinité d'inconnues.* Paris.
3. 1914. Demonstration nouvelle d'un théorème concernant les opérations fonctionelles linéaires. *Ann. Éc. Norm.* **31**, 9–14.

RIESZ, M.

1. 1922. Sur le problème des moments. Première Note. *Ark. för mat., astr. och fys.* **16**, no. 12 (23 pp.).
2. 1922. Deuxième Note. *ibid.* **16**, no. 19 (21 pp.).
3. 1923. Troisième Note. *ibid.* **17**, no. 16 (62 pp.).
4. 1922. Sur le problème des moments et le théorème de Parseval correspondant. *Acta Litt. ac Sci., Szeged.* **1**, 209–225.

SCHOENBERG, I. J.

1. 1938. Metric spaces and completely monotone functions. *Ann. of Math.* **39**, 811–841.
2. 1938. Metric spaces and positive definite functions. *Trans. Am. Math. Soc.* **44**, 522–536.

SCHUR, I.

1. 1911. Bemerkungen zur Theorie der beschränkten Bilinearformen mit unendlich vielen Veränderlichen. *J. f. Math.* **140**, 1–28.
2. 1917 & 1918. Über Potenzreihen, die im Inneren des Einheitskreises beschränkt sind. *J. f. Math.* **147**, 205–232; **148**, 122–145.

SIERPINSKI, W.

1. 1920. Sur les fonctions convexes mesurables. *Fund. Math.* (Warsaw) **1**, 125–129 (in French).

SHVETSOV, K. I.

1. 1939. On Hamburger's moment problem with the supplementary requirement that masses are absent on a given interval. *Commun. Soc. math. Kharkov* **16** (in Russian).

SONIN, N. YA.

1. 1954. The accuracy of determination of limiting values of integrals. *Collected Papers.* Moscow (in Russian).

BIBLIOGRAPHY

STIELTJES, T.

1. 1884. Quelques recherches sur la théorie des quadratures dites méchaniques. *Ann. Éc. Norm.* **1**, 409–426.
2. 1885. Note à l'occasion de la réclamation de M. Markoff. *Ann. Éc. Norm.* **2**, 183–4.
3. 1894–1895. Recherches sur les fractions continues. *Anns. Fac. Sci. Univ. Toulouse*, **8**, J1–J122 (page numbers); **9** A5–A47 (page numbers).

STONE, M.

1. 1932. *Linear transformations in Hilbert space and their applications to analysis.* Am. Math. Soc. Colloquium Publications, New York. vol. **15** (622 pp.).

SZEGÖ, G.

1. 1939. *Orthogonal Polynomials.* Amer. Math. Soc., New York.

SZEGÖ, G. & GRENANDER, U.

1. 1958. *Toeplitz Forms and their Applications.* Univ. Calif. Press.

TAMARKIN, J. D. & SHOHAT, J. A.

1. 1943. *The problem of moments.* Amer. Math. Soc., New York.

TITCHMARSH, E. C.

1. 1950. *Introduction to the theory of Fourier integrals*, 2nd Ed. O.U.P. London

TÖPLITZ, O.

1. 1911. Über die Fouriersche Entwicklung positiver Funktionen. *Rend. Circ. mat. Palermo* **32**, 191–192.

WALL, H.

1. 1948. *Analytic theory of continued fractions.* van Nostrand Co., New York.

WEYL, H.

1. 1910. Über gewöhnliche Differentialgeleichungen mit Singularitäten und die zugehörigen Entwicklungen willkürlicher Funktionen, *Math. Ann.* **68**, 220–269.
2. 1935. Über das Pick-Nevanlinnasche Interpolationsproblem und sein infinitesimales Analogon. *Ann. of Math.* **36**, 230–254.

WIDDER, D. V.

1. 1941. *The Laplace transform.* Princeton Univ. Press, New Jersey.

WOUK, A.

1. 1953. Difference equations and *J*-matrices. *Duke Math. J.* **20**, 141–159.

INDEX

Mathematics-Bestsellers

HANDBOOK OF MATHEMATICAL FUNCTIONS: with Formulas, Graphs, and Mathematical Tables, Edited by Milton Abramowitz and Irene A. Stegun. A classic resource for working with special functions, standard trig, and exponential logarithmic definitions and extensions, it features 29 sets of tables, some to as high as 20 places. 1046pp. 8 x 10 1/2.   0-486-61272-4

ABSTRACT AND CONCRETE CATEGORIES: The Joy of Cats, Jiri Adamek, Horst Herrlich, and George E. Strecker. This up-to-date introductory treatment employs category theory to explore the theory of structures. Its unique approach stresses concrete categories and presents a systematic view of factorization structures. Numerous examples. 1990 edition, updated 2004. 528pp. 6 1/8 x 9 1/4.   0-486-46934-4

MATHEMATICS: Its Content, Methods and Meaning, A. D. Aleksandrov, A. N. Kolmogorov, and M. A. Lavrent'ev. Major survey offers comprehensive, coherent discussions of analytic geometry, algebra, differential equations, calculus of variations, functions of a complex variable, prime numbers, linear and non-Euclidean geometry, topology, functional analysis, more. 1963 edition. 1120pp. 5 3/8 x 8 1/2.   0-486-40916-3

INTRODUCTION TO VECTORS AND TENSORS: Second Edition--Two Volumes Bound as One, Ray M. Bowen and C.-C. Wang. Convenient single-volume compilation of two texts offers both introduction and in-depth survey. Geared toward engineering and science students rather than mathematicians, it focuses on physics and engineering applications. 1976 edition. 560pp. 6 1/2 x 9 1/4.   0-486-46914-X

AN INTRODUCTION TO ORTHOGONAL POLYNOMIALS, Theodore S. Chihara. Concise introduction covers general elementary theory, including the representation theorem and distribution functions, continued fractions and chain sequences, the recurrence formula, special functions, and some specific systems. 1978 edition. 272pp. 5 3/8 x 8 1/2.
0-486-47929-3

ADVANCED MATHEMATICS FOR ENGINEERS AND SCIENTISTS, Paul DuChateau. This primary text and supplemental reference focuses on linear algebra, calculus, and ordinary differential equations. Additional topics include partial differential equations and approximation methods. Includes solved problems. 1992 edition. 400pp. 7 1/2 x 9 1/4.   0-486-47930-7

PARTIAL DIFFERENTIAL EQUATIONS FOR SCIENTISTS AND ENGINEERS, Stanley J. Farlow. Practical text shows how to formulate and solve partial differential equations. Coverage of diffusion-type problems, hyperbolic-type problems, elliptic-type problems, numerical and approximate methods. Solution guide available upon request. 1982 edition. 414pp. 6 1/8 x 9 1/4.   0-486-67620-X

VARIATIONAL PRINCIPLES AND FREE-BOUNDARY PROBLEMS, Avner Friedman. Advanced graduate-level text examines variational methods in partial differential equations and illustrates their applications to free-boundary problems. Features detailed statements of standard theory of elliptic and parabolic operators. 1982 edition. 720pp. 6 1/8 x 9 1/4.   0-486-47853-X

LINEAR ANALYSIS AND REPRESENTATION THEORY, Steven A. Gaal. Unified treatment covers topics from the theory of operators and operator algebras on Hilbert spaces; integration and representation theory for topological groups; and the theory of Lie algebras, Lie groups, and transform groups. 1973 edition. 704pp. 6 1/8 x 9 1/4.
0-486-47851-3

A SURVEY OF INDUSTRIAL MATHEMATICS, Charles R. MacCluer. Students learn how to solve problems they'll encounter in their professional lives with this concise single-volume treatment. It employs MATLAB and other strategies to explore typical industrial problems. 2000 edition. 384pp. 5 3/8 x 8 1/2.   0-486-47702-9

NUMBER SYSTEMS AND THE FOUNDATIONS OF ANALYSIS, Elliott Mendelson. Geared toward undergraduate and beginning graduate students, this study explores natural numbers, integers, rational numbers, real numbers, and complex numbers. Numerous exercises and appendixes supplement the text. 1973 edition. 368pp. 5 3/8 x 8 1/2.   0-486-45792-3

A FIRST LOOK AT NUMERICAL FUNCTIONAL ANALYSIS, W. W. Sawyer. Text by renowned educator shows how problems in numerical analysis lead to concepts of functional analysis. Topics include Banach and Hilbert spaces, contraction mappings, convergence, differentiation and integration, and Euclidean space. 1978 edition. 208pp. 5 3/8 x 8 1/2.   0-486-47882-3

FRACTALS, CHAOS, POWER LAWS: Minutes from an Infinite Paradise, Manfred Schroeder. A fascinating exploration of the connections between chaos theory, physics, biology, and mathematics, this book abounds in award-winning computer graphics, optical illusions, and games that clarify memorable insights into self-similarity. 1992 edition. 448pp. 6 1/8 x 9 1/4.   0-486-47204-3

SET THEORY AND THE CONTINUUM PROBLEM, Raymond M. Smullyan and Melvin Fitting. A lucid, elegant, and complete survey of set theory, this three-part treatment explores axiomatic set theory, the consistency of the continuum hypothesis, and forcing and independence results. 1996 edition. 336pp. 6 x 9.   0-486-47484-4

DYNAMICAL SYSTEMS, Shlomo Sternberg. A pioneer in the field of dynamical systems discusses one-dimensional dynamics, differential equations, random walks, iterated function systems, symbolic dynamics, and Markov chains. Supplementary materials include PowerPoint slides and MATLAB exercises. 2010 edition. 272pp. 6 1/8 x 9 1/4.   0-486-47705-3

ORDINARY DIFFERENTIAL EQUATIONS, Morris Tenenbaum and Harry Pollard. Skillfully organized introductory text examines origin of differential equations, then defines basic terms and outlines general solution of a differential equation. Explores integrating factors; dilution and accretion problems; Laplace Transforms; Newton's Interpolation Formulas, more. 818pp. 5 3/8 x 8 1/2.   0-486-64940-7

MATROID THEORY, D. J. A. Welsh. Text by a noted expert describes standard examples and investigation results, using elementary proofs to develop basic matroid properties before advancing to a more sophisticated treatment. Includes numerous exercises. 1976 edition. 448pp. 5 3/8 x 8 1/2.   0-486-47439-9

THE CONCEPT OF A RIEMANN SURFACE, Hermann Weyl. This classic on the general history of functions combines function theory and geometry, forming the basis of the modern approach to analysis, geometry, and topology. 1955 edition. 208pp. 5 3/8 x 8 1/2.   0-486-47004-0

THE LAPLACE TRANSFORM, David Vernon Widder. This volume focuses on the Laplace and Stieltjes transforms, offering a highly theoretical treatment. Topics include fundamental formulas, the moment problem, monotonic functions, and Tauberian theorems. 1941 edition. 416pp. 5 3/8 x 8 1/2.   0-486-47755-X

Mathematics–Logic and Problem Solving

PERPLEXING PUZZLES AND TANTALIZING TEASERS, Martin Gardner. Ninety-three riddles, mazes, illusions, tricky questions, word and picture puzzles, and other challenges offer hours of entertainment for youngsters. Filled with rib-tickling drawings. Solutions. 224pp. 5 3/8 x 8 1/2.   0-486-25637-5

MY BEST MATHEMATICAL AND LOGIC PUZZLES, Martin Gardner. The noted expert selects 70 of his favorite "short" puzzles. Includes The Returning Explorer, The Mutilated Chessboard, Scrambled Box Tops, and dozens more. Complete solutions included. 96pp. 5 3/8 x 8 1/2.   0-486-28152-3

THE LADY OR THE TIGER?: and Other Logic Puzzles, Raymond M. Smullyan. Created by a renowned puzzle master, these whimsically themed challenges involve paradoxes about probability, time, and change; metapuzzles; and self-referentiality. Nineteen chapters advance in difficulty from relatively simple to highly complex. 1982 edition. 240pp. 5 3/8 x 8 1/2.   0-486-47027-X

SATAN, CANTOR AND INFINITY: Mind-Boggling Puzzles, Raymond M. Smullyan. A renowned mathematician tells stories of knights and knaves in an entertaining look at the logical precepts behind infinity, probability, time, and change. Requires a strong background in mathematics. Complete solutions. 288pp. 5 3/8 x 8 1/2.
0-486-47036-9

THE RED BOOK OF MATHEMATICAL PROBLEMS, Kenneth S. Williams and Kenneth Hardy. Handy compilation of 100 practice problems, hints and solutions indispensable for students preparing for the William Lowell Putnam and other mathematical competitions. Preface to the First Edition. Sources. 1988 edition. 192pp. 5 3/8 x 8 1/2.   0-486-69415-1

KING ARTHUR IN SEARCH OF HIS DOG AND OTHER CURIOUS PUZZLES, Raymond M. Smullyan. This fanciful, original collection for readers of all ages features arithmetic puzzles, logic problems related to crime detection, and logic and arithmetic puzzles involving King Arthur and his Dogs of the Round Table. 160pp. 5 3/8 x 8 1/2.
0-486-47435-6

UNDECIDABLE THEORIES: Studies in Logic and the Foundation of Mathematics, Alfred Tarski in collaboration with Andrzej Mostowski and Raphael M. Robinson. This well-known book by the famed logician consists of three treatises: "A General Method in Proofs of Undecidability," "Undecidability and Essential Undecidability in Mathematics," and "Undecidability of the Elementary Theory of Groups." 1953 edition. 112pp. 5 3/8 x 8 1/2.   0-486-47703-7

LOGIC FOR MATHEMATICIANS, J. Barkley Rosser. Examination of essential topics and theorems assumes no background in logic. "Undoubtedly a major addition to the literature of mathematical logic." – *Bulletin of the American Mathematical Society.* 1978 edition. 592pp. 6 1/8 x 9 1/4.   0-486-46898-4

INTRODUCTION TO PROOF IN ABSTRACT MATHEMATICS, Andrew Wohlgemuth. This undergraduate text teaches students what constitutes an acceptable proof, and it develops their ability to do proofs of routine problems as well as those requiring creative insights. 1990 edition. 384pp. 6 1/2 x 9 1/4.   0-486-47854-8

FIRST COURSE IN MATHEMATICAL LOGIC, Patrick Suppes and Shirley Hill. Rigorous introduction is simple enough in presentation and context for wide range of students. Symbolizing sentences; logical inference; truth and validity; truth tables; terms, predicates, universal quantifiers; universal specification and laws of identity; more. 288pp. 5 3/8 x 8 1/2.   0-486-42259-3

Mathematics–Algebra and Calculus

VECTOR CALCULUS, Peter Baxandall and Hans Liebeck. This introductory text offers a rigorous, comprehensive treatment. Classical theorems of vector calculus are amply illustrated with figures, worked examples, physical applications, and exercises with hints and answers. 1986 edition. 560pp. 5 3/8 x 8 1/2. 0-486-46620-5

ADVANCED CALCULUS: An Introduction to Classical Analysis, Louis Brand. A course in analysis that focuses on the functions of a real variable, this text introduces the basic concepts in their simplest setting and illustrates its teachings with numerous examples, theorems, and proofs. 1955 edition. 592pp. 5 3/8 x 8 1/2. 0-486-44548-8

ADVANCED CALCULUS, Avner Friedman. Intended for students who have already completed a one-year course in elementary calculus, this two-part treatment advances from functions of one variable to those of several variables. Solutions. 1971 edition. 432pp. 5 3/8 x 8 1/2. 0-486-45795-8

METHODS OF MATHEMATICS APPLIED TO CALCULUS, PROBABILITY, AND STATISTICS, Richard W. Hamming. This 4-part treatment begins with algebra and analytic geometry and proceeds to an exploration of the calculus of algebraic functions and transcendental functions and applications. 1985 edition. Includes 310 figures and 18 tables. 880pp. 6 1/2 x 9 1/4. 0-486-43945-3

BASIC ALGEBRA I: Second Edition, Nathan Jacobson. A classic text and standard reference for a generation, this volume covers all undergraduate algebra topics, including groups, rings, modules, Galois theory, polynomials, linear algebra, and associative algebra. 1985 edition. 528pp. 6 1/8 x 9 1/4. 0-486-47189-6

BASIC ALGEBRA II: Second Edition, Nathan Jacobson. This classic text and standard reference comprises all subjects of a first-year graduate-level course, including in-depth coverage of groups and polynomials and extensive use of categories and functors. 1989 edition. 704pp. 6 1/8 x 9 1/4. 0-486-47187-X

CALCULUS: An Intuitive and Physical Approach (Second Edition), Morris Kline. Application-oriented introduction relates the subject as closely as possible to science with explorations of the derivative; differentiation and integration of the powers of x; theorems on differentiation, antidifferentiation; the chain rule; trigonometric functions; more. Examples. 1967 edition. 960pp. 6 1/2 x 9 1/4. 0-486-40453-6

ABSTRACT ALGEBRA AND SOLUTION BY RADICALS, John E. Maxfield and Margaret W. Maxfield. Accessible advanced undergraduate-level text starts with groups, rings, fields, and polynomials and advances to Galois theory, radicals and roots of unity, and solution by radicals. Numerous examples, illustrations, exercises, appendixes. 1971 edition. 224pp. 6 1/8 x 9 1/4. 0-486-47723-1

AN INTRODUCTION TO THE THEORY OF LINEAR SPACES, Georgi E. Shilov. Translated by Richard A. Silverman. Introductory treatment offers a clear exposition of algebra, geometry, and analysis as parts of an integrated whole rather than separate subjects. Numerous examples illustrate many different fields, and problems include hints or answers. 1961 edition. 320pp. 5 3/8 x 8 1/2. 0-486-63070-6

LINEAR ALGEBRA, Georgi E. Shilov. Covers determinants, linear spaces, systems of linear equations, linear functions of a vector argument, coordinate transformations, the canonical form of the matrix of a linear operator, bilinear and quadratic forms, and more. 387pp. 5 3/8 x 8 1/2. 0-486-63518-X

CATALOG OF DOVER BOOKS

Mathematics–Probability and Statistics

BASIC PROBABILITY THEORY, Robert B. Ash. This text emphasizes the probabilistic way of thinking, rather than measure-theoretic concepts. Geared toward advanced undergraduates and graduate students, it features solutions to some of the problems. 1970 edition. 352pp. 5 3/8 x 8 1/2.　　　　　　　　　　0-486-46628-0

PRINCIPLES OF STATISTICS, M. G. Bulmer. Concise description of classical statistics, from basic dice probabilities to modern regression analysis. Equal stress on theory and applications. Moderate difficulty; only basic calculus required. Includes problems with answers. 252pp. 5 5/8 x 8 1/4.　　　　　　　　　　0-486-63760-3

OUTLINE OF BASIC STATISTICS: Dictionary and Formulas, John E. Freund and Frank J. Williams. Handy guide includes a 70-page outline of essential statistical formulas covering grouped and ungrouped data, finite populations, probability, and more, plus over 1,000 clear, concise definitions of statistical terms. 1966 edition. 208pp. 5 3/8 x 8 1/2.　　　　　　　　　　0-486-47769-X

GOOD THINKING: The Foundations of Probability and Its Applications, Irving J. Good. This in-depth treatment of probability theory by a famous British statistician explores Keynesian principles and surveys such topics as Bayesian rationality, corroboration, hypothesis testing, and mathematical tools for induction and simplicity. 1983 edition. 352pp. 5 3/8 x 8 1/2.　　　　　　　　　　0-486-47438-0

INTRODUCTION TO PROBABILITY THEORY WITH CONTEMPORARY APPLICATIONS, Lester L. Helms. Extensive discussions and clear examples, written in plain language, expose students to the rules and methods of probability. Exercises foster problem-solving skills, and all problems feature step-by-step solutions. 1997 edition. 368pp. 6 1/2 x 9 1/4.　　　　　　　　　　0-486-47418-6

CHANCE, LUCK, AND STATISTICS, Horace C. Levinson. In simple, non-technical language, this volume explores the fundamentals governing chance and applies them to sports, government, and business. "Clear and lively ... remarkably accurate." – Scientific Monthly. 384pp. 5 3/8 x 8 1/2.　　　　　　　　　　0-486-41997-5

FIFTY CHALLENGING PROBLEMS IN PROBABILITY WITH SOLUTIONS, Frederick Mosteller. Remarkable puzzlers, graded in difficulty, illustrate elementary and advanced aspects of probability. These problems were selected for originality, general interest, or because they demonstrate valuable techniques. Also includes detailed solutions. 88pp. 5 3/8 x 8 1/2.　　　　　　　　　　0-486-65355-2

EXPERIMENTAL STATISTICS, Mary Gibbons Natrella. A handbook for those seeking engineering information and quantitative data for designing, developing, constructing, and testing equipment. Covers the planning of experiments, the analyzing of extreme-value data; and more. 1966 edition. Index. Includes 52 figures and 76 tables. 560pp. 8 3/8 x 11.　　　　　　　　　　0-486-43937-2

STOCHASTIC MODELING: Analysis and Simulation, Barry L. Nelson. Coherent introduction to techniques also offers a guide to the mathematical, numerical, and simulation tools of systems analysis. Includes formulation of models, analysis, and interpretation of results. 1995 edition. 336pp. 6 1/8 x 9 1/4.　　　　　　　　　　0-486-47770-3

INTRODUCTION TO BIOSTATISTICS: Second Edition, Robert R. Sokal and F. James Rohlf. Suitable for undergraduates with a minimal background in mathematics, this introduction ranges from descriptive statistics to fundamental distributions and the testing of hypotheses. Includes numerous worked-out problems and examples. 1987 edition. 384pp. 6 1/8 x 9 1/4.　　　　　　　　　　0-486-46961-1

Browse over 9,000 books at www.doverpublications.com

Mathematics–Geometry and Topology

PROBLEMS AND SOLUTIONS IN EUCLIDEAN GEOMETRY, M. N. Aref and William Wernick. Based on classical principles, this book is intended for a second course in Euclidean geometry and can be used as a refresher. More than 200 problems include hints and solutions. 1968 edition. 272pp. 5 3/8 x 8 1/2.   0-486-47720-7

TOPOLOGY OF 3-MANIFOLDS AND RELATED TOPICS, Edited by M. K. Fort, Jr. With a New Introduction by Daniel Silver. Summaries and full reports from a 1961 conference discuss decompositions and subsets of 3-space; n-manifolds; knot theory; the Poincaré conjecture; and periodic maps and isotopies. Familiarity with algebraic topology required. 1962 edition. 272pp. 6 1/8 x 9 1/4.   0-486-47753-3

POINT SET TOPOLOGY, Steven A. Gaal. Suitable for a complete course in topology, this text also functions as a self-contained treatment for independent study. Additional enrichment materials make it equally valuable as a reference. 1964 edition. 336pp. 5 3/8 x 8 1/2.   0-486-47222-1

INVITATION TO GEOMETRY, Z. A. Melzak. Intended for students of many different backgrounds with only a modest knowledge of mathematics, this text features self-contained chapters that can be adapted to several types of geometry courses. 1983 edition. 240pp. 5 3/8 x 8 1/2.   0-486-46626-4

TOPOLOGY AND GEOMETRY FOR PHYSICISTS, Charles Nash and Siddhartha Sen. Written by physicists for physics students, this text assumes no detailed background in topology or geometry. Topics include differential forms, homotopy, homology, cohomology, fiber bundles, connection and covariant derivatives, and Morse theory. 1983 edition. 320pp. 5 3/8 x 8 1/2.   0-486-47852-1

BEYOND GEOMETRY: Classic Papers from Riemann to Einstein, Edited with an Introduction and Notes by Peter Pesic. This is the only English-language collection of these 8 accessible essays. They trace seminal ideas about the foundations of geometry that led to Einstein's general theory of relativity. 224pp. 6 1/8 x 9 1/4. 0-486-45350-2

GEOMETRY FROM EUCLID TO KNOTS, Saul Stahl. This text provides a historical perspective on plane geometry and covers non-neutral Euclidean geometry, circles and regular polygons, projective geometry, symmetries, inversions, informal topology, and more. Includes 1,000 practice problems. Solutions available. 2003 edition. 480pp. 6 1/8 x 9 1/4.   0-486-47459-3

TOPOLOGICAL VECTOR SPACES, DISTRIBUTIONS AND KERNELS, François Trèves. Extending beyond the boundaries of Hilbert and Banach space theory, this text focuses on key aspects of functional analysis, particularly in regard to solving partial differential equations. 1967 edition. 592pp. 5 3/8 x 8 1/2.
     0-486-45352-9

INTRODUCTION TO PROJECTIVE GEOMETRY, C. R. Wylie, Jr. This introductory volume offers strong reinforcement for its teachings, with detailed examples and numerous theorems, proofs, and exercises, plus complete answers to all odd-numbered end-of-chapter problems. 1970 edition. 576pp. 6 1/8 x 9 1/4.   0-486-46895-X

FOUNDATIONS OF GEOMETRY, C. R. Wylie, Jr. Geared toward students preparing to teach high school mathematics, this text explores the principles of Euclidean and non-Euclidean geometry and covers both generalities and specifics of the axiomatic method. 1964 edition. 352pp. 6 x 9.   0-486-47214-0

Mathematics-History

THE WORKS OF ARCHIMEDES, Archimedes. Translated by Sir Thomas Heath. Complete works of ancient geometer feature such topics as the famous problems of the ratio of the areas of a cylinder and an inscribed sphere; the properties of conoids, spheroids, and spirals; more. 326pp. 5 3/8 x 8 1/2.　　　　　0-486-42084-1

THE HISTORICAL ROOTS OF ELEMENTARY MATHEMATICS, Lucas N. H. Bunt, Phillip S. Jones, and Jack D. Bedient. Exciting, hands-on approach to understanding fundamental underpinnings of modern arithmetic, algebra, geometry and number systems examines their origins in early Egyptian, Babylonian, and Greek sources. 336pp. 5 3/8 x 8 1/2.　　　　　0-486-25563-8

THE THIRTEEN BOOKS OF EUCLID'S ELEMENTS, Euclid. Contains complete English text of all 13 books of the Elements plus critical apparatus analyzing each definition, postulate, and proposition in great detail. Covers textual and linguistic matters; mathematical analyses of Euclid's ideas; classical, medieval, Renaissance and modern commentators; refutations, supports, extrapolations, reinterpretations and historical notes. 995 figures. Total of 1,425pp. All books 5 3/8 x 8 1/2.

Vol. I: 443pp.　0-486-60088-2
Vol. II: 464pp.　0-486-60089-0
Vol. III: 546pp.　0-486-60090-4

A HISTORY OF GREEK MATHEMATICS, Sir Thomas Heath. This authoritative two-volume set that covers the essentials of mathematics and features every landmark innovation and every important figure, including Euclid, Apollonius, and others. 5 3/8 x 8 1/2.

Vol. I: 461pp.　0-486-24073-8
Vol. II: 597pp.　0-486-24074-6

A MANUAL OF GREEK MATHEMATICS, Sir Thomas L. Heath. This concise but thorough history encompasses the enduring contributions of the ancient Greek mathematicians whose works form the basis of most modern mathematics. Discusses Pythagorean arithmetic, Plato, Euclid, more. 1931 edition. 576pp. 5 3/8 x 8 1/2.

0-486-43231-9

CHINESE MATHEMATICS IN THE THIRTEENTH CENTURY, Ulrich Libbrecht. An exploration of the 13th-century mathematician Ch'in, this fascinating book combines what is known of the mathematician's life with a history of his only extant work, the Shu-shu chiu-chang. 1973 edition. 592pp. 5 3/8 x 8 1/2.

0-486-44619-0

PHILOSOPHY OF MATHEMATICS AND DEDUCTIVE STRUCTURE IN EUCLID'S ELEMENTS, Ian Mueller. This text provides an understanding of the classical Greek conception of mathematics as expressed in Euclid's Elements. It focuses on philosophical, foundational, and logical questions and features helpful appendixes. 400pp. 6 1/2 x 9 1/4.　　　　　0-486-45300-6

BEYOND GEOMETRY: Classic Papers from Riemann to Einstein, Edited with an Introduction and Notes by Peter Pesic. This is the only English-language collection of these 8 accessible essays. They trace seminal ideas about the foundations of geometry that led to Einstein's general theory of relativity. 224pp. 6 1/8 x 9 1/4.　0-486-45350-2

HISTORY OF MATHEMATICS, David E. Smith. Two-volume history – from Egyptian papyri and medieval maps to modern graphs and diagrams. Non-technical chronological survey with thousands of biographical notes, critical evaluations, and contemporary opinions on over 1,100 mathematicians. 5 3/8 x 8 1/2.

Vol. I: 618pp.　0-486-20429-4
Vol. II: 736pp.　0-486-20430-8